Calculus 2 Review in Bite-Size Pieces

By Kathryn Paulk
Copyright © 2023

Updated: 06/01/2024

Table of Contents

Introduction

This book is a review for students who are currently taking or have already taken a second course in calculus.

Calculus 2 topics are presented in short bite-size pieces and/or short bite-size examples.

This book has been formatted so that it is easy to read on both paperback and also on electronic devices with the Kindle app (laptop, iPad, Kindle E-reader, and iPhone).

Calculus 2 Review

A short review of Calculus 2 topics are included in the following sections. Proofs are not included. Most theorems are in summary form. Examples are included to reinforce the topics.

Integration by Parts (IBP)

Integration By Parts (IBP) — Equation

If $\quad u = f(x) \quad$ and $\quad v = g(x)$

then

$$\int u\, dv \;=\; uv - \int v\, du$$

In other words ...

If you find it difficult to integrate, scrambling it up may help!

IBP -- Picking Parts with LIATE

Integration By Parts (IBP)
Use "LIATE" to Pick the Parts

$$\int u\,dv \;=\; uv - \int v\,du$$

The first function listed in the acronym "LIATE" is the u part.

The rest is the dv

L	=	Log Functions
I	=	Inverse Trig Functions
A	=	Algebraic Functions
T	=	Trig Functions
E	=	Exponential Functions

Integration By Parts (IBP)   − Ex. 2a

Evaluate: $I_1 = \int t^2\, e^t\, dt$

Use LI$\underline{A}$TE to find $\boldsymbol{u}$	$\boldsymbol{u = t^2}$ $\frac{du}{dt} = 2t$ $du = 2t\, dt$	$dv = e^t dt$ $\int dv = \int e^t dt$ $v = e^t$

IBP: $\int u\; dv = \; uv - \int v\; du$

$I_1 = \int t^2\, e^t\, dt \;=\; t^2 e^t - \int e^t\, 2t\, dt$

$I_1 = \; t^2 e^t - \int e^t\, 2t\, dt$

$I_1 = \; t^2 e^t - 2\int e^t\, t\, dt$

$I_1 = \; t^2 e^t - 2\, I_2$

Use IBP again to solve: $I_2 = \int e^t\, t\, dt$
Continued …

Integration By Parts (IBP) — Ex. 2b

Evaluate: $I_2 = \int e^t\, t\, dt$

Use LIA̲TE to find u	$u = t$ $\frac{du}{dt} = 1$ $du = dt$	$dv = e^t\, dt$ $\int dv = \int e^t\, dt$ $v = e^t$

IBP: $\int u\, dv = uv - \int v\, du$

$I_2 = \int t\, e^t\, dt = te^t - \int e^t\, dt$

$I_2 = te^t - e^t$

$I_1 =$ Original Integral

$I_1 = t^2 e^t - 2 \cdot I_2$

$I_1 = t^2 e^t - 2[\, te^t - e^t\,]$

$I_1 = t^2 e^t - 2te^t + 2e^t \quad + \quad C$

Integration By Parts (IBP) — Ex. 3

Evaluate: $I = \int x \sin x \, dx$

Use LIATE to find u	$u = x$ $\frac{du}{dx} = 1$ $du = dx$	$dv = \sin x \, dx$ $\int dv = \int \sin x \, dx$ $v = -\cos x$

IBP: $\int u \, dv = uv - \int v \, du$

$I = x(-\cos x) - \int(-\cos x)\, dx$

$I = -x\cos x + \sin x$

$I = -x\cos x + \sin x + C$

Integration By Parts (IBP) — Ex. 4

Evaluate: $I = \int \ln x \, dx$

Use	$u = \ln x$	$dv = dx$
LIATE to find **u**	$\frac{du}{dx} = \frac{1}{x}$	$\int dv = \int dx$
	$du = \frac{1}{x} dx$	$v = x$

IBP: $\int u \, dv = uv - \int v \, du$

$I = uv - \int v \, du$

$I = \ln x \cdot x - \int x \left(\frac{1}{x} dx\right)$

$I = \ln x \cdot x - \int dx$

$I = x \ln x - x + C$

IBP -- Tabular Method

Integration By Parts Tabular Method is similar to regular IBP but is faster and more efficient to use.

	IBP Tabular Method — Process (1/2)
	$$\int u\,dv \ = \ uv - \int v\,du$$
1.	Use the acronym "LIATE" to identify the u part. The rest is the dv part.
2.	Create a table, with u at the top of the 1st column. For each row in the 1st column, keep taking the derivative until one of the following happens: • The derivative $= 0$. • You can't take the derivative. • u was a log function, so just stop after taking the first derivative. • The derivative $= n \cdot u$, a multiple of the original u part.
3.	Continued …

	IBP Tabular Method – Process (2/2)
	Continued …
3.	Put the dv part at the top of the 2nd column. For each row in the 2nd column, keep taking the integral.
4.	Between the two columns, put alternating $+/-$ signs. Start with $+$
5.	Draw short, diagonal arrows connecting rows in 1st column to one lower row in the 2nd column.
6.	In last row draw a horizontal arrow.
7.	Write the answer by following the arrows. Multiply diagonal arrows and add or subtract, based on the sign. For horizontal arrows, take the integral.

IBP Tabular Method − Ex. 1

Evaluate: $I = \int (2x)\cos(9x)\,dx$

Use L I <u>A</u> T E to find u. Setup table.

Du		$\int dv$
$2x$	$+$	$\cos(9x)\,dx$
2	$-$	$\dfrac{1}{9}\sin(9x)$
0	$+$	$\dfrac{-1}{81}\cos(9x)$
Take derivative of this side until you can't or it's $= 0$		Keep taking integral of this side.

$$I = 2x\left(\tfrac{1}{9}\right)\sin 9x + \tfrac{2}{81}\cos 9x + \int 0\,dx$$

Note: Last horizontal arrow ➔ Integral

IBP Tabular Method — Ex. 2

Evaluate: $I = \int t^2 e^t \, dt$

Du	$\int dv$
$t^2 \quad +$	$e^t \, dt$
$2t \quad -$	e^t
$2 \quad +$	e^t
$0 \quad -$	e^t
Take derivative of this side until you can't or it's $= 0$	Keep taking integral of this side.

$I = t^2 e^t - 2te^t + 2e^t - \int e^t \cdot 0 \, dx$

$I = t^2 e^t - 2te^t + 2e^t + C$

Note: Last horizontal arrow → Integral

IBP Tabular Method – Ex. 3

Evaluate: $I = \int x^5 \ln x \; dx$

Du		$\int dv$
$\ln x$	$+$	$x^5 dx$
$\dfrac{1}{x}$	$-$	$\dfrac{x^6}{6}$
Stop here with logs.		

$I = \ln x \left(\dfrac{1}{6}\right) x^6 - \int \left(\dfrac{1}{6}\right) x^5 \, dx$

$I = \dfrac{1}{6} \left[x^6 \ln x - \int x^5 \, dx \right]$

$I = \dfrac{1}{6} \left[x^6 \ln x - \dfrac{x^6}{6} \right]$

$I = \dfrac{x^6 \ln x}{6} - \dfrac{x^6}{36} + C$

IBP Tabular Method — Ex. 4

Evaluate: $I = \int e^{2x} \cos 3x \; dx$

Du		$\int dv$
$\cos 3x$	$+$	$e^{2x} dx$
$-\,3\sin 3x$	$-$	$\left(\dfrac{1}{2}\right) e^{2x}$
$-\,9\cos 3x$	$+$	$\left(\dfrac{1}{4}\right) e^{2x}$
Stop with $\;n \cdot \boldsymbol{u}$		

$$I = \frac{(\cos 3x)\, e^{2x}}{2} + \frac{3(\sin 3x)\, e^{2x}}{4}$$

$$+ \int \left(\frac{-9}{4}\right)(\cos 3x)e^{2x} dx$$

$$I = \frac{(\cos 3x)\, e^{2x}}{2} + \frac{3(\sin 3x)\, e^{2x}}{4} - \left(\frac{9}{4}\right) I$$

$$\left(\frac{13}{4}\right) I = \frac{(\cos 3x)\, e^{2x}}{2} + \frac{3(\sin 3x)\, e^{2x}}{4}$$

$$I = \left(\frac{4}{13}\right)\left[\frac{(\cos 3x)\, e^{2x}}{2} + \frac{3(\sin 3x)\, e^{2x}}{4} \right]$$

IBP Tabular Method — Ex. 5a

Evaluate: $I = \int x^4 e^{2x} \, dx$

Du		$\int dv$
x^4	$+$	$e^{2x} dx$
$4x^3$	$-$	$\left(\frac{1}{2}\right) e^{2x}$
$12x^2$	$+$	$\left(\frac{1}{4}\right) e^{2x}$
$24x$	$-$	$\left(\frac{1}{8}\right) e^{2x}$
24	$+$	$\left(\frac{1}{16}\right) e^{2x}$
0	$-$	$\left(\frac{1}{32}\right) e^{2x}$

$I =$ Continued ...

IBP Tabular Method — Ex. 5b

Evaluate: $I = \int x^4 e^{2x} \; dx$

Continued from previous page ...

$$I = \frac{x^4 e^{2x}}{2} - \frac{4x^3 e^{2x}}{4} + \frac{12x^2 e^{2x}}{8} \;\ldots$$

$$- \frac{24x\, e^{2x}}{16} + \frac{24\, e^{2x}}{32}$$

$$I = e^{2x}\left[\frac{x^4}{2} - \frac{4x^3}{4} + \frac{12x^2}{8} - \frac{24x}{16} + \frac{24}{32}\right]$$

$$I = e^{2x}\left[\frac{x^4}{2} - x^3 + \frac{3x^2}{2} - \frac{3x}{2} + \frac{3}{4}\right] + C$$

Trigonometric Integrals

Trig Integrals -- Guidelines		
$\int \sin^m x \cos^n x \; dx$		
m	n	Do This
---	Odd	Save one Cos for the dx
Odd	---	Save one Sin for the dx
Even	Even	Use Trig Identities

Some Useful Trig Identities
$\sin x \cos x = \frac{1}{2} \sin 2x$
$\sin 2x = 2 \sin x \cos x$
$\cos 2x = \cos^2 x - \sin^2 x$
$\sin^2 x = \frac{1}{2}(1 - \cos 2x)$
$\cos^2 x = \frac{1}{2}(1 + \cos 2x)$

Trig Integrals – Ex. 0
Evaluate: $I = \int \sin^2(x)\ dx$

Just use Trig. formulas.	$\sin^2 u = \frac{1}{2}(1 - \cos(2u))$
	$\cos^2 u = \frac{1}{2}(1 + \cos(2u))$
	$\sin(2u) = 2\sin(u)\cos(u)$

$$I = \int \frac{1}{2}(1 - \cos(2x)\ dx$$

$$I = \int \frac{1}{2}\ dx - \int \cos(2x)\ dx$$

$$\boxed{u = 2x \quad \rightarrow \quad du = 2\ dx}$$

$$I = \int \frac{1}{2}\ dx - \left(\frac{1}{2}\right)\left(\frac{1}{2}\right)\int \cos(2x)\ 2\ dx$$

$$I = \frac{x}{2} - \left(\frac{1}{4}\right)\sin(2x)$$

$$I = \frac{x}{2} - \left(\frac{1}{4}\right)[2\sin(x)\cdot\cos(x)]$$

$$I = \frac{x}{2} - \frac{\sin(x)\cdot\cos(x)}{2} + C$$

Trig Integrals – Ex. 1
Evaluate: $\int \cos^3 x \; dx$

Save one Cos for dx	$\int \cos^2 x \cdot \cos x \; dx$

$\int (1 - \sin^2 x) \cdot \cos x \; dx$

$\int (\cos x - \sin^2 x \cos x) \, dx$

$\int \cos x \; dx \; - \; \int \sin^2 x \cos x \, dx$

$\int \cos x \; dx \; - \; \int \sin^2 x \, (\cos x \, dx)$

$\sin x \; - \; \int u^2 \, du$

$\sin x \; - \; \frac{1}{3} u^3$

$\sin x \; - \; \frac{1}{3} \sin^3 x \; + \; C$

(Stewart, Calculus Early Transcendentals, p. 479)

Trig Integrals – Ex. 2	
Evaluate: $\int \sin^5 x \cos^2 x \ dx$	
Save one Sin for dx	$\int \sin^4 x \ \cos^2 x \cdot \sin x \, dx$

$\int (1 - \cos^2 x)^2 \cdot \cos^2 x \cdot \sin x \, dx$

Let: $u = \cos x, \quad du = -\sin x \, dx$

$\int (1 - u^2)^2 \, u^2 \cdot (-du)$

$-\int (1 - 2u^2 + u^4) \, u^2 \cdot (du)$

$-\int (u^2 - 2u^4 + u^6) \cdot du$

$-\left[\frac{1}{3}u^3 - \frac{2}{5}u^5 + \frac{1}{7}u^7 \right]$

$-\frac{1}{3}u^3 + \frac{2}{5}u^5 - \frac{1}{7}u^7$

$-\frac{1}{3}\cos^3 x + \frac{2}{5}\cos^5 x - \frac{1}{7}\cos^7 x + C$

(Stewart, Calculus Early Transcendentals, p. 479)

Trig Integrals – Ex. 3
Evaluate: $\int \sin^2 x \ dx$

Use Trig Identity	$\int \frac{1}{2}(1 - \cos 2x)\, dx$

$\frac{1}{2}\int (1 - \cos 2x)\, dx$

$\frac{1}{2}\int dx \ - \ \frac{1}{2}\int \cos 2x \ dx$

<u>Let</u>: $u = 2x$, $du = 2dx$

$\frac{1}{2}\int dx \ - \ \frac{1}{2}\cdot\frac{1}{2}\int \cos u\,(du)$

$\frac{1}{2}x \ - \ \frac{1}{4}\sin u$

$\frac{1}{2}x \ - \ \frac{1}{4}\sin 2x \ + \ C$

(Stewart, Calculus Early Transcendentals, p. 480)

Trig Integrals – Ex. 4

Evaluate: $\int sin^4 x \ dx$

Use Trig Identity	$\int \left[\frac{1}{2}(1 - \cos 2x)\right]^2 dx$ $\int \frac{1}{4}(1 - \cos 2x)^2 \ dx$

$$\frac{1}{4} \int (1 - 2\cos 2x + \cos^2 2x) \ dx$$

$$\frac{1}{4} \int \left[1 - 2\cos 2x + \frac{1}{2}(1 + \cos 4x)\right] dx$$

$$\frac{1}{4} \int \left[\frac{3}{2} - 2\cos 2x + \frac{1}{2}\cos 4x\right] dx$$

$$\frac{1}{4} \left[\frac{3}{2}x - \sin 2x + \frac{1}{8}\sin 4x\right] + C$$

(Stewart, Calculus Early Transcendentals, p. 480)

Trig Integrals – More Guidelines		
$\int tan^m\, x \, sec^n\, x \;\; dx$		
m	n	Do This
---	Even	Save one $sec^2 x$ for the dx
Odd	---	Save one $(sec\, x \tan x)$ for dx

Some trig derivatives & integrals
$\dfrac{d}{dx}\tan u = sec^2 u \cdot u'$ $\dfrac{d}{dx}\sec u = (\sec u \cdot \tan u)\, u'$
$\int \tan u\, du = \ln\|\sec u\| + C$ $\int \sec u\, du = \ln\|\sec u + \tan u\| + C$ $\int sec^2 u\, du = \tan u + C$ $\int sec^3 u\, du =$ $= \frac{1}{2}[\sec u \cdot \tan u + \ln\|\sec u + \tan u\|] + C$

Trig Integrals – Substituting tan & sec	
Equations for $\tan^2 x$ and $\sec^2 x$ are easy to derive.	
Pythagorean Identity	$\sin^2 x + \cos^2 x = 1$
Divide by $\cos^2 x$	$\dfrac{\sin^2 x}{\cos^2 x} + \dfrac{\cos^2 x}{\cos^2 x} = \dfrac{1}{\cos^2 x}$ $\tan^2 x + 1 = \sec^2 x$
Solve for $\tan^2 x$ & $\sec^2 x$	$\tan^2 x = \sec^2 x - 1$ $\sec^2 x = \tan^2 x + 1$

Trig Integrals – Ex. 5	
Evaluate: $\int tan^6 x \ sec^4 x \ dx$	
Separate one $sec^2 x$	$\int tan^6 x \ sec^2 x \ sec^2 x \ dx$

$$\int tan^6 x \ (1 - tan^2 x) \ sec^2 x \ dx$$

Let: $u = \tan x$ & $du = \sec^2 x \ dx$

$$\int u^6 \ (1 - u^2) \ du$$

$$\int (u^6 - u^8) \ du$$

$$\frac{u^7}{7} - \frac{u^9}{9}$$

$$\frac{\tan^7 x}{7} - \frac{\tan^9 x}{9} + C$$

(Stewart, Calculus Early Transcendentals, p. 481)

Trig Integrals – Ex. 6

Evaluate: $\int tan^5 x \; sec^7 x \; dx$

$\int tan^4 x \; sec^6 x \; (\sec x \tan x) \; dx$

$\int (sec^2 x - 1)^2 \; sec^6 x \; (\sec x \tan x) \; dx$

$\int (sec^4 x - 2 \, sec^2 x + 1) \, sec^6 x$
$$\cdot (\sec x \tan x) \; dx$$

$\int (sec^{10} x - 2 \, sec^8 x + sec^6 x)$
$$\cdot (\sec x \tan x) \; dx$$

$\int (u^{10} - 2u^8 + u^6) \cdot du$

$\dfrac{u^{11}}{11} - \dfrac{2u^9}{9} + \dfrac{u^7}{7}$

$\dfrac{sec^{11} x}{11} - \dfrac{2 \, sec^9 x}{9} + \dfrac{sec^7 x}{7} + C$

(Stewart, Calculus Early Transcendentals, p. 482)

Trig Integrals – Ex. 7
Evaluate: $\int tan^3 x \ dx$

$\int \tan x \cdot tan^2 x \ dx$

$\int \tan x \cdot (sec^2 x - 1) \ dx$

$\int (\tan x \sec^2 x - \tan x) \ dx$

$\int \tan x \ (\sec x \ dx) \ - \ \int \tan x \ dx$

$\int u \ (du) \ - \ \int \tan x \ dx$

$\frac{u^2}{2} - (-\ln|\cos x|)$

$\frac{u^2}{2} + \ln|\cos x|$

$\frac{tan^2 x}{2} + \ln|\cos x| \ + \ C$ (can stop here)

$\frac{tan^2 x}{2} + \ln| \ (\sec x)^{-1}| \ + \ C$

$\frac{tan^2 x}{2} - \ln| \sec x \ | \ + \ C$

(Stewart, Calculus Early Transcendentals, p. 483)

Trig Integrals – Ex. 8a

Evaluate: $\int \sec^3 x \; dx$

Note: Trig integrals, involving $\sec u$, often occur and it is a good idea to memorize the following equations.

$$\int \sec u \, du \;=\; \ln|\sec u + \tan u| \;+\; C$$

$$\int \sec^2 u \, du \;=\; \tan u \;+\; C$$

$$\int \sec^3 u \, du \;=$$

$$=\frac{1}{2}[\sec u \cdot \tan u \;+\; \ln|\sec u + \tan u|] + C$$

Solution:

$$\int \sec^3 x \; dx \;=$$

$$=\frac{1}{2}[\sec x \cdot \tan x \;+\; \ln|\sec x + \tan x|] + C$$

Continued ... (Without the above eqn.)

Trig Integrals – Ex. 8b
Evaluate: $I = \int sec^3 x \ dx$ $I = \int sec\, x \cdot sec^2 x \ dx$

Integrate by Parts Use: LI**A**T**E**	$u = sec\, x$ $du = sec\, x \tan x \ dx$
	$dv = sec^2 x \ dx$ $v = \tan x$
	$\int u \ dv = uv - \int v \ du$

$I = sec\, x \tan x - \int sec\, x \tan^2 x \ dx$

$I = sec\, x \tan x - \int sec\, x \,(sec^2 x - 1) \, dx$

$I = sec\, x \tan x - \int (sec^3 x - sec\, x) \, dx$

$I = sec\, x \tan x - \int sec^3 x \ dx \ + \int sec\, x \ dx$

$I = sec\, x \tan x \ - \ I \ + \int sec\, x \ dx$

$2I = sec\, x \tan x \ + \int sec\, x \ dx$

$2I = sec\, x \tan x \ + \ \ln|sec\, x + \tan x|$

$I = \frac{1}{2}[sec\, x \tan x \ + \ \ln|sec\, x + \tan x|] + C$

(Stewart, Calculus Early Transcendentals, p. 483)

Trig Integrals – More Guidelines

For integrals with
$$sin(mx) \quad and \quad cos(nx)$$

For integrals with $\sin A$ and $\cos B$
Use these trig identities.

$$\sin A \cos B = \frac{1}{2}[\sin(A - B) + \sin(A + B)]$$

$$\sin A \sin B = \frac{1}{2}[\cos(A - B) - \cos(A + B)]$$

$$\cos A \cos B = \frac{1}{2}[\cos(A - B) + \cos(A + B)]$$

35

Trig Integrals – Ex. 9

Evaluate: $\int \sin 4x \cos 5x \; dx$

$I = \int \sin 4x \cos 5x \; dx$

$I = \int \frac{1}{2}[\sin(-x) + \sin(9x)] \, dx$

$I = \frac{1}{2} \int [-\sin(x) + \sin(9x)] \, dx$

$I = \frac{1}{2} \left[-(-\cos x) + \frac{1}{9}\cos 9x \right]$

$I = \frac{1}{2} \left[\cos x + \frac{1}{9}\cos 9x \right] + C$

(Stewart, Calculus Early Transcendentals, p. 484)

Trigonometric Substitution

When an integral includes algebraic terms that are difficult to integrate, we may be able to substitute trig expressions to simplify the integral.

To use trig substitution, the original integral must have parts that can be represented by a right triangle.

Such as: $a,\ a^2,\ u,\ u^2,\ \sqrt{a^2 \pm u^2}$

Trig Substitution -- Examples

Example	Integral
1.	$\int \frac{\sqrt{9-x^2}}{x^2}\, dx$
2.	$\int \sqrt{a^2-x^2}\, dx$
3.	$\int \frac{1}{x^2\sqrt{x^2+4}}\, dx$
4.	$\int \frac{x}{\sqrt{x^2+4}}\, dx$
5.	$\int \frac{1}{\sqrt{x^2-a^2}}\, dx$
6.	$\int \frac{x^3}{(4x^2+9)^{\frac{3}{2}}}\, dx$
7.	$\int \frac{x}{\sqrt{3-2x-x^2}}\, dx$

(Stewart, Calculus Early Transcendentals, p. 486 - 491)

Trig Substitution -- Triangles

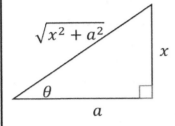

$$\tan \theta = \frac{x}{a}$$

$$x = a \cdot \tan \theta$$

$$dx = a \cdot \sec^2 \theta \, d\theta$$

$$\cos \theta = \frac{a}{\sqrt{x^2 + a^2}}$$

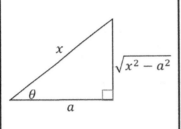

$$\cos \theta = \frac{a}{x}$$

$$x = a \cdot \cos \theta$$

$$dx = -a \cdot \sin \theta \, d\theta$$

$$\tan \theta = \frac{\sqrt{x^2 - a^2}}{a}$$

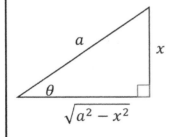

$$\sin \theta = \frac{x}{a}$$

$$x = a \cdot \sin \theta$$

$$dx = a \cdot \cos \theta \, d\theta$$

$$\cos \theta = \frac{\sqrt{a^2 - x^2}}{a}$$

Standard Trig Substitutions

$\sqrt{a^2 - x^2}$	$x = a \cdot \sin\theta, \quad -\frac{\pi}{2} \leq \theta \leq \frac{\pi}{2}$
	$1 - \sin^2\theta = \cos^2\theta$
$\sqrt{a^2 + x^2}$	$x = a\tan\theta, \quad -\frac{\pi}{2} \leq \theta \leq \frac{\pi}{2}$
	$1 + \tan^2\theta = \sec^2\theta$
$\sqrt{x^2 - a^2}$	$x = a\sec\theta, \quad 0 \leq \theta \leq \frac{\pi}{2}$ or $\quad \pi \leq \theta \leq \frac{3\pi}{2}$
	$\sec^2\theta - 1 = \tan^2\theta$

(Stewart, Calculus Early Transcendentals, p. 486)

There are two similar, but different, approaches for Trig Substitution.

The **standard trig substitutions** are based on the Pythagorean Theorem for a right triangle and the definition of trig functions.

The standard trig substitutions are often used, but I find using **substitutions, based on triangle sketches**, easier to do and easier to understand.

Either approach will work. In this section, the first problem will be solved twice – first using substitutions based on triangles, then again by using standard trig substitutions. All other problems will be solved using substitutions based on triangle sketches.

Trig Substitution – Ex. 1a

$$\text{Evaluate: } I = \int \frac{\sqrt{9-x^2}}{x^2}\, dx$$

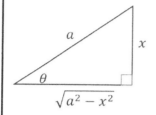

$\sin\theta = \dfrac{x}{a}$

$x = a \cdot \sin\theta$

$dx = a \cdot \cos\theta\, d\theta$

$\cos\theta = \dfrac{\sqrt{a^2-x^2}}{a}$

$\sqrt{a^2-x^2} = a \cdot \cos\theta$

$$I = \int \frac{3 \cdot \cos\theta}{(3\sin\theta)^2}\, 3\cos\theta\, d\theta$$

$$I = \int \frac{9 \cdot \cos^2\theta}{9 \cdot \sin^2\theta}\, d\theta = \int \cot^2\theta\, d\theta$$

$$I = \int (\csc^2\theta - 1)\, d\theta$$

$$I = -\cot\theta - \theta \quad \text{(from integral tables)}$$

$$I = -\frac{\sqrt{9-x^2}}{x} - \sin^{-1}\left(\frac{x}{3}\right) + C$$

Trig Substitution – Ex. 1b

Evaluate: $I = \int \dfrac{\sqrt{9-x^2}}{x^2}\,dx$

$\sqrt{a^2 - x^2}$	$x = a \cdot \sin\theta\,,\quad -\dfrac{\pi}{2} \le \theta \le \dfrac{\pi}{2}$
	$1 - \sin^2\theta = \cos^2\theta$

$I = \int \dfrac{\sqrt{9 - 9\sin^2\theta}}{(3\sin\theta)^2}\,3\cos\theta\,d\theta$

$I = \int \dfrac{3\cdot\cos\theta}{(3\sin\theta)^2}\,3\cos\theta\,d\theta$

$I = \int \dfrac{9\cdot\cos^2\theta}{9\cdot\sin^2\theta}\,d\theta \;=\; \int \cot^2\theta\,d\theta$

$I = \int (\csc^2\theta - 1)\,d\theta$

$I = -\cot\theta - \theta \quad$ (from integral tables)

$I = -\dfrac{\sqrt{9-x^2}}{x} - \sin^{-1}\left(\dfrac{x}{3}\right) + C$

(Stewart, Calculus Early Transcendentals, p. 486)

Trig Substitution – Ex. 2a

Find the area of ellipse: $\dfrac{x^2}{a^2} + \dfrac{y^2}{b^2} = 1$

Rewrite as a function of y	$\dfrac{y^2}{b^2} = 1 - \dfrac{x^2}{a^2}$
	$y^2 = b^2 \left(\dfrac{a^2 - x^2}{a^2}\right)$
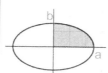	$y = \pm \dfrac{b}{a} \sqrt{a^2 - x^2}$

$A = Area$	$A = 4 \int_0^a \dfrac{b}{a} \sqrt{a^2 - x^2}\, dx$

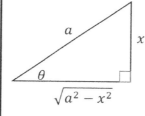	$\sin\theta = \dfrac{x}{a}$
	$x = a \cdot \sin\theta$
	$dx = a \cdot \cos\theta\, d\theta$
	$\cos\theta = \dfrac{\sqrt{a^2 - x^2}}{a}$
	$\sqrt{a^2 - x^2} = a \cdot \cos\theta$

Continued ...

Trig Substitution – Ex. 2b

$$A = 4 \left(\frac{b}{a}\right) \int_{x=0}^{x=a} \sqrt{a^2 - x^2}\, dx$$

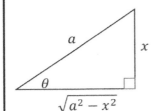

$$\sin\theta = \frac{x}{a}$$

$$x = a \cdot \sin\theta$$

$$dx = a \cdot \cos\theta\, d\theta$$

$$\cos\theta = \frac{\sqrt{a^2 - x^2}}{a}$$

$$\sqrt{a^2 - x^2} = a \cdot \cos\theta$$

$$A = 4 \left(\frac{b}{a}\right) \int_{\theta=0}^{\theta=\frac{\pi}{2}} a\cos\theta \,(a \cdot \cos\theta\, d\theta)$$

$$A = 4\,ab \int_0^{\frac{\pi}{2}} \cos^2\theta\, d\theta$$

$$A = 4\,ab \int_0^{\frac{\pi}{2}} \frac{1}{2}(1 + \cos 2\theta)\, d\theta$$

$$A = 2\,ab \int_0^{\frac{\pi}{2}} (1 + \cos 2\theta)\, d\theta$$

$$A = 2ab \left[\theta + \frac{1}{2}\sin 2\theta \right]_0^{\frac{\pi}{2}}$$

$$A = 2ab \left[\frac{\pi}{2} \right] = ab \cdot \pi$$

Trig Substitution – Ex. 3
Evaluate: $I = \int \dfrac{1}{x^2 \sqrt{x^2+4}}\, dx$

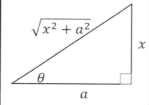

$$\tan\theta = \frac{x}{a}$$

$$x = a \cdot \tan\theta$$

$$dx = a \cdot \sec^2\theta\, d\theta$$

$$\cos\theta = \frac{a}{\sqrt{x^2+a^2}}$$

$$\sqrt{x^2+a^2} = \frac{a}{\cos\theta}$$

$$I = \int \frac{1}{(2\tan\theta)^2 \left(\frac{2}{\cos\theta}\right)} \,(2\sec^2\theta\, d\theta)$$

$$I = \int \frac{1}{\left(\frac{4\sin^2\theta}{\cos^2\theta}\right)\left(\frac{2}{\cos\theta}\right)} \left(\frac{2}{\cos^2\theta}\, d\theta\right)$$

$$I = \int \frac{\cos\theta}{4\sin^2\theta}\, d\theta \;=\; \frac{1}{4}\int \frac{1}{u^2}\, du$$

$$I = \frac{1}{4}\int u^{-2}\, du = \frac{1}{4}\left[(-1)u^{-1}\right] = -\frac{1}{4u}$$

$$I = -\frac{1}{4\sin\theta} = -\frac{1}{4\left(\frac{x}{\sqrt{x^2+4}}\right)}$$

$$I = -\frac{\sqrt{x^2+4}}{4x} + C$$

Trig Substitution – Ex. 4a

Evaluate: $I = \int \frac{x}{\sqrt{x^2 + 4}}\, dx$

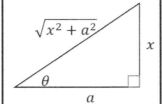

$\tan\theta = \frac{x}{a}$

$x = a \cdot \tan\theta$

$dx = a \cdot \sec^2\theta\, d\theta$

$\cos\theta = \frac{a}{\sqrt{x^2 + a^2}}$

$\sqrt{x^2 + a^2} = \frac{a}{\cos\theta}$

$I = \int \frac{2\tan\theta}{\left(\frac{2}{\cos\theta}\right)}\, (2\sec^2\theta\, d\theta)$

$I = \int \sin\theta \left(\frac{2}{\cos^2\theta}\, d\theta\right)$

$I = 2\int \frac{\sin\theta}{\cos^2\theta}\, d\theta \;=\; 2\int \frac{1}{u^2}\, du$

$I = 2\int u^{-2}\, du = 2\left[(-1)u^{-1}\right] = -\frac{2}{u}$

$I = -\frac{2}{\cos\theta} = -\frac{2}{\left(\frac{2}{\sqrt{x^2+4}}\right)}$

$I = -\sqrt{x^2+4} + C_1 \;=\; \sqrt{x^2+4} + C$

Trig Substitution – Ex. 4b

$$\text{Evaluate: } I \;=\; \int \frac{x}{\sqrt{x^2+4}}\; dx$$

Just use u-sub	$u = x^2 + 4$
This approach Is better!!!	$\frac{du}{dx} = 2x$ $du = 2x\, dx$

$$I \;=\; \left(\tfrac{1}{2}\right) \int \frac{1}{\sqrt{x^2+4}}\; 2x dx$$

$$I = \left(\tfrac{1}{2}\right) \int u^{-\frac{1}{2}}\; du$$

$$I = \left(\tfrac{1}{2}\right) \left[\frac{u^{\frac{1}{2}}}{\frac{1}{2}} \right] = u^{\frac{1}{2}}$$

$$I = \sqrt{x^2+4} \;+\; C$$

(Stewart, Calculus Early Transcendentals, p. 489)

Trig Substitution – Ex. 5

Evaluate: $I = \int \frac{1}{\sqrt{x^2 - a^2}}\, dx$

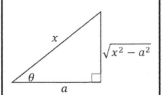

$\cos \theta = \dfrac{a}{x}$

$x = a \cdot \cos \theta$

$dx = -a \cdot \sin \theta\, d\theta$

$\tan \theta = \dfrac{\sqrt{x^2 - a^2}}{a}$

$\sqrt{x^2 - a^2} = a \cdot \tan \theta$

$I = \int \dfrac{1}{a \tan \theta}\ (-a \sin \theta\, d\theta)$

$I = \int -\dfrac{\sin^2 \theta}{\cos \theta}\, d\theta = \int -\dfrac{(1 - \cos^2 \theta)}{\cos \theta}\, d\theta$

$I = \int \dfrac{(\cos^2 \theta - 1)}{\cos \theta}\, d\theta = \int (\cos \theta - \sec \theta)\, d\theta$

$I = \sin \theta - \ln|\sec \theta + \tan \theta|$

$I = \dfrac{\sqrt{x^2 - a^2}}{x} - \ln\left|\dfrac{x}{a} + \dfrac{\sqrt{x^2 - a^2}}{a}\right|$

$I = \dfrac{\sqrt{x^2 - a^2}}{x} - \ln\left|x + \sqrt{x^2 - a^2}\right| + \ln a$

$I = \dfrac{\sqrt{x^2 - a^2}}{x} - \ln\left|x + \sqrt{x^2 - a^2}\right| + C$

Trig Substitution – Ex. 6a

Evaluate: $I = \int \dfrac{x^3}{(4x^2+9)^{\frac{3}{2}}}\, dx$

Rearrange	$I = \int \dfrac{x^3}{\left(4\left(x^2+\frac{9}{4}\right)\right)^{\frac{3}{2}}}\, dx$
Remove the coefficient of the x^2 term.	$I = \int \dfrac{x^3}{(4)^{\frac{3}{2}}\left(x^2+\frac{9}{4}\right)^{\frac{3}{2}}}\, dx$
	$I = \int \dfrac{x^3}{8\left(x^2+\frac{9}{4}\right)^{\frac{3}{2}}}\, dx$
	$I = \frac{1}{8}\int \left(\dfrac{x}{\sqrt{x^2+\frac{9}{4}}}\right)^3 dx$

Continued ...

50

Trig Substitution – Ex. 6b

Evaluate: $I = \frac{1}{8}\int \left(\dfrac{x}{\sqrt{x^2 + \frac{9}{4}}}\right)^3 dx$

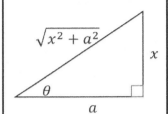

$\tan\theta = \dfrac{x}{a}$

$x = a\cdot\tan\theta$

$dx = a\cdot\sec^2\theta\, d\theta$

$\cos\theta = \dfrac{a}{\sqrt{x^2 + a^2}}$

$\sqrt{x^2 + a^2} = \dfrac{a}{\cos\theta}$

$I = \frac{1}{8}\int \left(\dfrac{a\tan\theta}{\frac{a}{\cos\theta}}\right)^3 (a\,\sec^2\theta\,d\theta) \quad ; \quad a = \frac{3}{2}$

$I = (a)\frac{1}{8}\int (\tan\theta\cdot\cos\theta)^3 \left(\dfrac{1}{\cos^2\theta}\right) d\theta$

$I = \frac{3}{16}\int \tan^3\theta\cdot\cos\theta\, d\theta$

$I = \frac{3}{16}\int \dfrac{\sin^3\theta}{\cos^2\theta} d\theta = \frac{3}{16}\int \dfrac{\sin^2\theta}{\cos^2\theta}\sin\theta\, d\theta$

Continued ...

Trig Substitution – Ex. 6c

Evaluate: $I = \frac{3}{16} \int \frac{\sin^2 \theta}{\cos^2 \theta} \sin \theta \, d\theta$

$I = \frac{3}{16} \int \frac{1 - \cos^2 \theta}{\cos^2 \theta} \sin \theta \, d\theta$

$I = \frac{3}{16} \int (\cos^{-2} \theta - 1) \sin \theta \, d\theta$

$I = \frac{3}{16} \left[\int -u^{-2} du - \int \sin \theta \, d\theta \right]$

$I = -\frac{3}{16} \left[\int u^{-2} du + \int \sin \theta \, d\theta \right]$

$I = -\frac{3}{16} \left[-u^{-1} - \cos \theta \right]$

$I = \frac{3}{16} \left[\frac{1}{\cos \theta} + \cos \theta \right]$

$I = \frac{3}{16} \left[\frac{1}{\cos \theta} + \cos \theta \right] = \frac{3}{16} \left[\frac{1 + \cos^2 \theta}{\cos \theta} \right]$

$I = \frac{3}{16} \left[\frac{1 + 1 - \sin^2 \theta}{\cos \theta} \right] = \frac{3}{16} \left[\frac{2 - \sin^2 \theta}{\cos \theta} \right]$

Continued ...

Trig Substitution – Ex. 6d

Evaluate: $I = \dfrac{3}{16}\left[\dfrac{2 - \sin^2\theta}{\cos\theta}\right]$

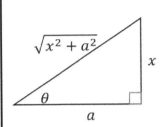

$\tan\theta = \dfrac{x}{a}$

$x = a \cdot \tan\theta$

$dx = a \cdot \sec^2\theta\, d\theta$

$\cos\theta = \dfrac{a}{\sqrt{x^2 + a^2}}$

$\sqrt{x^2 + a^2} = \dfrac{a}{\cos\theta}$

$I = \dfrac{3}{16}\left[\dfrac{2 - \left(\dfrac{x}{\sqrt{x^2 + a^2}}\right)^2}{\left(\dfrac{a}{\sqrt{x^2 + a^2}}\right)}\right] \qquad a = \dfrac{3}{2}$

$I = \left(\dfrac{1}{a}\right)\dfrac{3}{16}\left[2\sqrt{x^2 + a^2} - \dfrac{x^2}{\sqrt{x^2 + a^2}}\right]$

$I = \dfrac{1}{8}\left[\dfrac{2(x^2 + a^2) - x^2}{\sqrt{x^2 + a^2}}\right] = \dfrac{1}{8}\left[\dfrac{x^2 + 2a^2}{\sqrt{x^2 + a^2}}\right]$

$I = \dfrac{1}{8}\left[\dfrac{x^2 + 2\left(\dfrac{9}{4}\right)}{\sqrt{x^2 + \dfrac{9}{4}}}\right] = \dfrac{1}{8}\left[\dfrac{x^2 + 2\left(\dfrac{9}{4}\right)}{\sqrt{x^2 + \dfrac{9}{4}}} \cdot \left(\dfrac{2}{2}\right)\right]$

$I = \dfrac{2x^2 + 9}{8\sqrt{4x^2 + 9}}$

Trig Substitution – Ex. 7a

Evaluate: $I = \int \dfrac{x}{\sqrt{3 - 2x - x^2}}\, dx$

Complete the square to change the form to: $\sqrt{a^2 - u^2}$	$3 - 2x - x^2$
	$3 - (2x + x^2)$
	$3 - (x^2 + 2x + 1) + 1$
	$4 - (x + 1)^2$
	$4 - u^2$

Evaluate: $I = \int \dfrac{u - 1}{\sqrt{4 - u^2}}\, dx$

Where: $u = x + 1 \ ; \quad x = u - 1$

$du = dx$

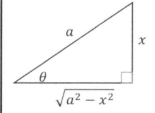

$\sin \theta = \dfrac{x}{a}$

$x = a \cdot \sin \theta$

$dx = a \cdot \cos \theta\, d\theta$

$\cos \theta = \dfrac{\sqrt{a^2 - x^2}}{a}$

$\sqrt{a^2 - x^2} = \dfrac{a}{\cos \theta}$

Continued ...

Trig Substitution – Ex. 7b

Evaluate: $I = \int \dfrac{u-1}{\sqrt{4-u^2}}\, du$

Where: $u = x + 1 \;\; ; \;\; x = u - 1$

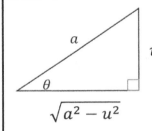

$\sin\theta = \dfrac{u}{a}$

$u = a \cdot \sin\theta$

$du = a \cdot \cos\theta\, d\theta$

$\cos\theta = \dfrac{\sqrt{a^2 - u^2}}{a}$

$\sqrt{a^2 - u^2} = a \cdot \cos\theta$

$I = \int \dfrac{u-1}{\sqrt{4-u^2}}\, du \qquad\qquad ; \; a = 2$

$I = \int \dfrac{a\sin\theta - 1}{a\cos\theta}\, a\cos\theta\, d\theta$

$I = \int (a\sin\theta - 1)\, d\theta \;=\; -a\cos\theta\, - \, \theta$

$I = -\sqrt{a^2 - u^2} \, - \, \sin^{-1}\left(\dfrac{u}{a}\right)$

$I = -\sqrt{2^2 - (x+1)^2} \, - \, \sin^{-1}\left(\dfrac{x+1}{2}\right)$

$I = -\sqrt{3 - x^2 - 2x} \, - \, \sin^{-1}\left(\dfrac{x+1}{2}\right) + C$

Trig Substitution – Ex. 8

Evaluate: $I = \int \sqrt{1 - x^2} \; dx$

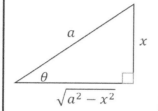

$\sin \theta = \dfrac{x}{a}$

$x = a \cdot \sin \theta$

$dx = a \cdot \cos \theta \; d\theta$

$\cos \theta = \dfrac{\sqrt{a^2 - x^2}}{a}$

$\sqrt{a^2 - x^2} = a \cdot \cos \theta$

$I = \int (a \cdot \cos \theta) \; (a \cdot \cos \theta \; d\theta) \qquad a = 1$

$I = \int \cos^2 \theta \; d\theta = \int \dfrac{1}{2}[1 + \cos(2\theta)] \; d\theta$

$I = \dfrac{1}{2} \int [1 - \cos(2\theta)] \; d\theta \qquad u = 2\theta$

$I = \dfrac{1}{2} \left[\theta - \dfrac{1}{2}\sin(2\theta) \right]$

$I = \dfrac{1}{2} \left[\theta - \dfrac{1}{2}(2 \sin \theta \cdot \cos \theta) \right]$

$I = \dfrac{1}{2} \left[\sin^{-1}(x) - (x)\left(\sqrt{1 - x^2}\right) \right] + C$

Integrating Rational Functions by Partial Fractions

Using Partial Fractions is an algebraic process used to simplify rational functions. This process is helpful when integrating rational functions.

Partial Fractions -- Overview	
The goal is to convert a complex rational function into several simpler rational functions. This makes integration easier. This is the general idea: $I = \int \left(\frac{R(x)}{Q(x)} \right) dx = \int \left(\frac{A}{x} + \frac{B}{x} \right) dx$ $I = A \cdot \ln|x| + B \cdot \ln|x|$	
Case I	The denominator, $Q(x)$, is a product of distinct linear factors.
Case II	The denominator, $Q(x)$, is a product of linear factors, some of which are repeated.
Case III	$Q(x)$ has irreducible quadratic factors, none of which repeat.
Case IV	$Q(x)$ contains a repeated, irreducible quadratic factor.

Partial Fractions -- Process

Step	Rational Function $= \dfrac{R(x)}{Q(x)}$
1.	If the degree of numerator function, $R(x)$, is greater than the degree of the denominator function, $Q(x)$, use polynomial division. The remainder will be a simpler, rational function.
2.	Factor the denominator, $Q(x)$.
3.	Using guidelines for the particular case, convert the rational function to a set of partial fractions. $$\frac{R(x)}{Q(x)} = \frac{A}{a(x)} + \frac{B}{b(x)} \dots$$
4.	Use algebra to solve for A, B, C ...
5.	Solve the Integral. $$\int \frac{R(x)}{Q(x)}\,dx = \int \left(\frac{A}{a(x)} + \frac{B}{b(x)} \dots \right) dx$$

Partial Fractions – Case I

The denominator of the rational function, $\frac{R(x)}{Q(x)}$, is a product of distinct linear factors.

$$\frac{R(x)}{Q(x)} = \frac{R(x)}{a(x) \cdot b(x)} = \frac{A}{a(x)} + \frac{B}{b(x)}$$

$$R(x) = [a(x) \cdot b(x)] \cdot \left[\frac{A}{a(x)} + \frac{B}{b(x)} \right]$$

$$R(x) = A \cdot b(x) + B \cdot a(x)$$

Solve for A and B

Partial Fractions: Case 1 – Ex. 1

$$\text{Evaluate: } I = \int \frac{x^3 + x}{x - 2}\, dx$$

Note: If degree of numerator is greater than degree of denominator, use long or synthetic division.

$$
\begin{array}{c|cccc}
 & x^3 & x^2 & x^1 & x^0 \\
\hline
2 & 1 & 0 & 1 & 0 \\
 & \downarrow & 2 & 4 & 10 \\
\hline
 & 1 & 2 & 5 & 10 \\
\end{array}
$$

$$I = \int \left(x^2 + 2x + 5 + \frac{10}{x-2} \right) dx$$

$$I = \int \left(x^2 + 2x + 5 + \frac{10}{x-2} \right) dx$$

$$I = \frac{x^3}{3} + 2\frac{x^2}{2} + 5x + 10 \cdot \ln|x-2|$$

$$I = \frac{x^3}{3} + x^2 + 5x + 10 \cdot \ln|x-2| + C$$

Partial Fractions: Case 1 – Ex. 2a

Evaluate: $I = \int \dfrac{x^2 + 2x - 3}{2x^3 + 3x^2 - 2x}\, dx$

Use long or synthetic division?	Degree of denominator is greater than degree of numerator so no need to divide. Just factor.

$$\frac{x^2 + 2x - 3}{2x^3 + 3x^2 - 2x} = \frac{x^2 + 2x - 3}{x(2x^2 + 3x - 2)}$$

$$= \frac{x^2 + 2x - 3}{x(2x - 1)(x + 2)} = \frac{A}{x} + \frac{B}{(2x - 1)} + \frac{C}{(x + 2)}$$

$x^2 + 2x - 3$

$= A(2x - 1)(x + 2) + B(x)(x + 2) + C(x)(2x - 1)$

$x = 0$	$-3 = A(-1)(2)$
	$-3 = -2A \quad \rightarrow \quad A = \frac{3}{2}$

$x = -2$	$-3 = C(-2)(-5)$
	$-3 = 10C \quad \rightarrow \quad C = \frac{-3}{10}$

Continued …

Partial Fractions: Case 1 – Ex. 2b

Evaluate: $I = \int \dfrac{x^2 + 2x - 3}{2x^3 + 3x^2 - 2x}\, dx$

$x^2 + 2x - 3$
$= A(2x - 1)(x + 2) + B(x)(x + 2) + C(x)(2x - 1)$

Previously, we found: $A = \dfrac{3}{2}$ and $C = \dfrac{-3}{10}$

$x = \dfrac{1}{2}$

$-\dfrac{7}{4} = B\left(\dfrac{1}{2}\right)\left(\dfrac{1}{2} + 2\right)$

$-\dfrac{7}{4} = B\left(\dfrac{1}{2}\right)\left(\dfrac{5}{2}\right) = B\left(\dfrac{5}{4}\right)$

$-7 = 5B \qquad \rightarrow B = -\dfrac{7}{5}$

$I = \int \left[\dfrac{x^2 + 2x - 3}{x(2x - 1)(x + 2)}\right] dx$

$I = \int \left[\dfrac{A}{x} + \dfrac{B}{(2x-1)} + \dfrac{C}{(x+2)}\right] dx$

$u = 2x - 1$
$du = 2dx$

$I = \dfrac{3}{2}\ln|x| - \dfrac{7}{5}\left(\dfrac{1}{2}\right)\ln|2x - 1| - \dfrac{3}{10}\ln|x + 2| + C$

$I = \dfrac{3}{2}\ln|x| - \dfrac{7}{10}\ln|2x - 1| - \dfrac{3}{10}\ln|x + 2| + C$

Partial Fractions: Case 1 – Ex. 3

$$\text{Evaluate: } I = \int \frac{3}{x^2 - a^2}\, dx$$

$$\frac{3}{x^2 - a^2} = \frac{3}{(x-a)(x+a)}$$

$$= \frac{3}{(x-a)(x+a)} = \frac{A}{(x-a)} + \frac{B}{(x+a)}$$

$$3 = A(x+a) + B(x-a)$$

$x = a$	$3 = A(2a)$	$\rightarrow A = \frac{3}{2a}$
$x = -a$	$3 = B(-2a)$	$\rightarrow B = -\frac{3}{2a}$

$$I = \int \left[\frac{3}{x^2-a^2}\right] dx = \int \left[\frac{A}{(x-a)} + \frac{B}{(x+a)}\right] dx$$

$$I = \int \left[\frac{\frac{3}{2a}}{(x-a)} - \frac{\frac{3}{2a}}{(x+a)}\right] dx$$

$$I = \frac{3}{2a}\ln|x-a| - \frac{3}{2a}\ln|x+a|$$

$$I = \frac{3}{2a}\left(\ln|x-a| - \ln|x+a|\right)$$

$$I = \frac{3}{2a}\left(\ln\frac{|x-a|}{|x+a|}\right) = \frac{3}{2a}\left(\ln\left|\frac{x-a}{x+a}\right|\right) + C$$

Partial Fractions – Case II

The denominator of the rational function, $\frac{R(x)}{Q(x)}$, is a product of linear factors,

some of which are repeated.

$$\frac{R(x)}{Q(x)} = \frac{R(x)}{h(x) \cdot h(x) \cdot k(x)} = \frac{A}{h(x)} + \frac{B}{\left(h(x)\right)^2} + \frac{C}{k(x)}$$

$$R(x) =$$
$$= \left[\left(h(x)\right)^2 \cdot k(x) \right] \cdot \left[\frac{A}{h(x)} + \frac{B}{\left(h(x)\right)^2} + \frac{C}{k(x)} \right]$$

$$R(x) = A \cdot h(x) \cdot k(x)$$
$$+ B \cdot k(x) + C \cdot \left(h(x)\right)^2$$

Solve for A, B, C

Partial Fractions: Case 2 – Ex. 1a

Evaluate: $I = \int \dfrac{x^4 - 2x^2 + 4x + 1}{x^3 - x^2 - x + 1} \, dx$

Long division...

$$
\begin{array}{r}
x + 1 \\
x^3 - x^2 - x + 1 \overline{\big)\; x^4 + 0x^3 - 2x^2 + 4x + 1} \\
-(x^4 - x^3 - x^2 + x\;) \\
\hline
x^3 - x^2 + 3x + 1 \\
-(x^3 - x^2 - x + 1\;) \\
\hline
4x
\end{array}
$$

$$\frac{x^4 - 2x^2 + 4x + 1}{x^3 - x^2 - x + 1} = x + 1 + \frac{4x}{x^3 - x^2 - x + 1}$$

Factor the denominator...   Hint: Factor by grouping

$$
\begin{aligned}
x^3 - x^2 - x + 1 &= (x^3 - x^2) - (x - 1) \\
&= x^2(x - 1) - (x - 1) \\
&= (x - 1)(x^2 - 1) \\
&= (x - 1)(x - 1)(x + 1) \\
&= (x - 1)^2(x + 1)
\end{aligned}
$$

$$I = \int \left[\frac{x^4 - 2x^2 + 4x + 1}{x^3 - x^2 - x + 1} \right] dx$$

$$I = \int \left[x + 1 + \frac{4x}{(x-1)^2(x+1)} \right] dx$$

Continued ...

66

Partial Fractions: Case 2 – Ex. 1b

Evaluate: $I = \int \left[x + 1 + \dfrac{4x}{(x-1)^2(x+1)} \right] dx$

$\dfrac{4x}{(x-1)^2(x+1)} = \dfrac{A}{(x-1)} + \dfrac{B}{(x-1)^2} + \dfrac{C}{(x+1)}$

$4x = A(x-1)(x+1) + B(x+1) + C(x-1)^2$

$x = 1$	$4 = B(2) \qquad \rightarrow B = 2$
$x = -1$	$-4 = C(4) \qquad \rightarrow C = -1$

$x = 0$	$0 = A(-1)(1) + B(1) + C(1)$
	$0 = -A + B$
	$0 = -A + 2 \qquad$ Recall: $B = 2$
	$A = 2$

$I = \int \left[x + 1 + \dfrac{4x}{(x-1)^2(x+1)} \right] dx$

$I = \int \left[x + 1 + \dfrac{A}{(x-1)} + \dfrac{B}{(x-1)^2} + \dfrac{C}{(x+1)} \right] dx$

$I = \int \left[x + 1 + \dfrac{2}{(x-1)} + \dfrac{2}{(x-1)^2} - \dfrac{1}{(x+1)} \right] dx$

$I = \dfrac{x^2}{2} + x + 2\ln|x-1| + \int \dfrac{2}{(x-1)^2} - \ln|x+1|$

Continued ...

Partial Fractions: Case 2 – Ex. 1c

$I = \int \left[x + 1 + \frac{4x}{(x-1)^2(x+1)} \right] dx$

$I = \int \left[x + 1 + \frac{A}{(x-1)} + \frac{B}{(x-1)^2} + \frac{C}{(x+1)} \right] dx$

$I = \int \left[x + 1 + \frac{2}{(x-1)} + \frac{2}{(x-1)^2} - \frac{1}{(x+1)} \right] dx$

$I = \frac{x^2}{2} + x + 2\ln|x-1| + \int \frac{2}{(x-1)^2} - \ln|x+1|$

$\int \frac{2}{(x-1)^2}$	$= 2\int (x-1)^{-2} \, dx$	$u = (x-1)$
		$du = dx$
	$= 2\frac{(x-1)^{-1}}{-1} = \frac{-2}{x-1}$	

$I = \frac{x^2}{2} + x + 2\ln|x-1| - \frac{2}{x-1} - \ln|x+1| + C$

Partial Fractions – Case III

The denominator of the rational function, $\frac{R(x)}{Q(x)}$, has irreducible quadratic factors, none of which repeat.

$$\frac{R(x)}{Q(x)} = \frac{R(x)}{h(x) \cdot k(x^2)} = \frac{A}{h(x)} + \frac{Bx + C}{k(x^2)}$$

$$R(x) =$$
$$= [h(x) \cdot k(x^2)] \cdot \left[\frac{A}{h(x)} + \frac{Bx + C}{k(x^2)}\right]$$

$$R(x) = A \cdot k(x^2) + (Bx + C) \cdot h(x)$$

Solve for A, B, C

Partial Fractions: Case 3 – Ex. 1

Evaluate: $I = \int \dfrac{1}{x^2 + 6x + 11}\, dx$

When the denominator is in the form:
$$ax^2 + bx + c$$

If possible, complete the square to get:

$$\int \frac{dx}{x^2 + a^2} = \frac{1}{a}\tan^{-1}\left(\frac{x}{a}\right) + C$$

$I = \int \dfrac{1}{x^2 + 6x + 11}\, dx$

$I = \int \dfrac{1}{(x^2 + 6x + 9) + 11 - 9}\, dx$

$I = \int \dfrac{1}{(x+3)^2 + 2}\, dx$

$I = \int \dfrac{1}{(x+3)^2 + \left(\sqrt{2}\right)^2}\, dx$

$I = \frac{1}{\sqrt{2}}\tan^{-1}\left(\frac{x+3}{\sqrt{2}}\right) + C$

$I = \frac{\sqrt{2}}{2}\tan^{-1}\left(\frac{\sqrt{2}\,(x+3)}{2}\right) + C$

Rationalize the denominators.

<u>Conclusion</u>: If you rearrange the integral with completing the square, using partial fractions may not be necessary. Instead, just use:

$$\int \frac{dx}{x^2 + a^2} = \frac{1}{a}\tan^{-1}\left(\frac{x}{a}\right) + C$$

Partial Fractions: Case 3 – Ex. 2a

Evaluate: $I = \int \dfrac{2x^2 - x - 8}{x^3 + 4x}\, dx$

$\dfrac{2x^2 - x - 8}{x^3 + 4x} = \dfrac{2x^2 - x - 8}{x(x^2 + 4)} = \dfrac{A}{x} + \dfrac{Bx + C}{(x^2 + 4)}$

$2x^2 - x - 8 = A(x^2 + 4) + (Bx + C)(x)$

$x = 0$	$-8 = A(4)$ $\qquad\qquad \to\; A = -2$
$x = 1$	$2 - 1 - 8 = A(5) + (B + C)(1)$ $-7 = 5A + B + C$ $-7 = 5(-2) + B + C$ $-7 = -10 + B + C$ $3 = B + C \qquad\qquad \to\; B = 3 - C$
$x = -1$	$2 + 1 - 8 = A(5) + (-B + C)(-1)$ $-5 = 5A + B - C$ $-5 = -10 + B - C$ $5 = B - C \qquad\qquad \to\; B = 5 + C$
$B = B$	$B = B$ $3 - C = 5 + C$ $-2 = 2C \qquad\qquad \to C = -1$ $\qquad\qquad\qquad \to B = 4$

$I = \int \left[\dfrac{2x^2 - x - 8}{x(x^2 + 4)}\right] dx = \int \left[\dfrac{A}{x} + \dfrac{Bx + C}{(x^2 + 4)}\right] dx$

Continued ...

Partial Fractions: Case 3 – Ex. 2b

$$I = \int \left[\frac{2x^2 - x - 8}{x(x^2 + 4)} \right] dx = \int \left[\frac{A}{x} + \frac{Bx + C}{(x^2 + 4)} \right] dx$$

Previously Found	$A = -2, \ B = 4, \ C = -1$

$$\frac{2x^2 - x - 8}{x^3 + 4x} = \frac{2x^2 - x - 8}{x(x^2 + 4)} = \frac{A}{x} + \frac{Bx + C}{(x^2 + 4)}$$

$$I = \int \left[-\frac{2}{x} + \frac{4x - 1}{(x^2 + 4)} \right] dx$$

$$I = \int \left[-\frac{2}{x} + 2\frac{2x}{(x^2 + 4)} - \frac{1}{(x^2 + 4)} \right] dx$$

$$I = -2ln|x| + 2\ln|x^2 + 4| - \int \frac{1}{(x^2 + 4)} dx$$

$$\int \frac{du}{a^2 + u^2} = \frac{1}{a}\tan^{-1}\left(\frac{u}{a}\right) + C$$

$$I = -2ln|x| + 2\ln(x^2 + 4) - \frac{1}{2}\tan^{-1}\left(\frac{x}{2}\right)$$

$$I = 2[\ln(x^2 + 4) - ln|x|] - \frac{1}{2}\tan^{-1}\left(\frac{x}{2}\right)$$

$$I = 2\ln\left|\frac{x^2 + 4}{x}\right| - \frac{1}{2}\tan^{-1}\left(\frac{x}{2}\right) + C$$

72

Partial Fractions: Case 3 – Ex. 2c

Evaluate: $I = \int \dfrac{2x^2 - x - 8}{x^3 + 4x}\, dx$

$$\frac{2x^2 - x - 8}{x^3 + 4x} = \frac{2x^2 - x - 8}{x(x^2 + 4)} = \frac{A}{x} + \frac{Bx + C}{(x^2 + 4)}$$

Another way to solve for $A, B,$ and C

$$2x^2 - x - 8 = A(x^2 + 4) + (Bx + C)(x)$$
$$2x^2 - x - 8 = Ax^2 + 4A + Bx^2 + Cx$$
$$2x^2 - x - 8 = (A + B)x^2 + (C)x + (4A)$$

Match coefficients	$A + B = 2$ $C = -1$ $4A = -8 \quad \to A = -2$ $\to B = 4$

$$A = -2,\ B = 4,\ C = -1$$

Partial Fractions: Case 3 – Ex. 3a

Evaluate: $I = \int \dfrac{4x^2 - 3x + 4}{4x^2 - 4x + 3}\, dx$

Degree of denominator is not less than degree of numerator so use long division...

$$
\begin{array}{r}
1 \\
4x^2 - 4x + 3 \enclose{longdiv}{4x^2 - 3x + 4} \\
-(4x^2 - 4x + 3) \\
\hline
x + 1
\end{array}
$$

$$\frac{4x^2 - 3x + 4}{4x^2 - 4x + 3} = 1 + \frac{x + 1}{4x^2 - 4x + 3}$$

The denominator cannot be factored because the discriminant is negative.

$$b^2 - 4ac = 4^2 - 4(4)(3) = -48$$

So, we cannot use the partial fraction technique. Instead, complete the square in the denominator.

Then, use: $\int \dfrac{dx}{x^2 + a^2} = \dfrac{1}{a}\tan^{-1}\left(\dfrac{x}{a}\right) + C$

$I = \int \left[\dfrac{4x^2 - 3x + 4}{4x^2 - 4x + 3}\right] dx$

$I = \int \left[1 + \dfrac{x + 1}{4x^2 - 4x + 3}\right] dx$

Continued ...

Partial Fractions: Case 3 – Ex. 3b

Evaluate: $I = \int \frac{4x^2 - 3x + 4}{4x^2 - 4x + 3} \, dx$

Previously We found	$I = \int \left[1 + \frac{x+1}{4x^2 - 4x + 3}\right] dx$
The plan	Complete the square in the denominator. Then use: $\int \frac{dx}{x^2 + a^2} = \frac{1}{a}\tan^{-1}\left(\frac{x}{a}\right) + C$

$I = \int \left[1 + \frac{x+1}{(4x^2 - 4x + 1) + 3 - 1}\right] dx$

$I = \int \left[1 + \frac{x+1}{(2x-1)^2 + 2}\right] dx$

$I = \int \left[1 + \frac{x}{(2x-1)^2 + 2} + \frac{1}{(2x-1)^2 + 2}\right] dx$

$I = \int 1\, dx + \int \frac{x\, dx}{(2x-1)^2 + 2} + \int \frac{dx}{(2x-1)^2 + 2}$

$I = x + \int \frac{x\, dx}{(2x-1)^2 + 2} + \left[\frac{1}{\sqrt{2}}\tan^{-1}\left(\frac{x}{\sqrt{2}}\right)\right]$

Continued ...

Partial Fractions: Case 3 – Ex. 3c

Evaluate: $I = \int \dfrac{4x^2 - 3x + 4}{4x^2 - 4x + 3}\, dx$

Previously we found...

$I = x + \int \dfrac{x\, dx}{(2x - 1)^2 + 2} + \left[\dfrac{1}{\sqrt{2}} \tan^{-1}\left(\dfrac{x}{\sqrt{2}} \right) \right]$

Use u-sub	$u = 2x - 1 \quad \rightarrow \quad x = \dfrac{u + 1}{2}$ $\dfrac{du}{dx} = 2 \qquad \rightarrow \quad dx = \dfrac{du}{2}$

$\displaystyle \int \frac{x\, dx}{(2x - 1)^2 + 2} = \int \frac{\left(\frac{u+1}{2}\right)\left(\frac{du}{2}\right)}{u^2 + 2}$

$= \left(\dfrac{1}{4}\right) \int \dfrac{u + 1}{u^2 + 2}\, du$

$= \left(\dfrac{1}{4}\right) \int \left[\dfrac{u}{u^2 + 2} + \dfrac{1}{u^2 + 2} \right] du$

$= \left(\dfrac{1}{4}\right) \left[\dfrac{1}{2}\ln(u^2 + 2) + \dfrac{1}{\sqrt{2}}\tan^{-1}\left(\dfrac{u}{\sqrt{2}} \right) \right]$

$= \dfrac{1}{8}\ln(u^2 + 2) + \dfrac{1}{4\sqrt{2}}\tan^{-1}\left(\dfrac{u}{\sqrt{2}} \right)$

$= \dfrac{1}{8}\ln((2x - 1)^2 + 2) + \dfrac{1}{4\sqrt{2}}\tan^{-1}\left(\dfrac{2x-1}{\sqrt{2}} \right)$

$I = x + \dfrac{1}{8}\ln((2x - 1)^2 + 2) + \dfrac{1}{4\sqrt{2}}\tan^{-1}\left(\dfrac{2x - 1}{\sqrt{2}} \right)$
$+ \left[\dfrac{1}{\sqrt{2}}\tan^{-1}\left(\dfrac{x}{\sqrt{2}} \right) \right] + C$

Partial Fractions – Case IV
The denominator of the rational function, $\frac{R(x)}{Q(x)}$, has contains an irreducible quadratic factor that repeats.
$\frac{R(x)}{Q(x)} = \frac{R(x)}{\left(h(x^2)\right)^3} = \frac{Ax+B}{h(x^2)} + \frac{Cx+D}{\left(h(x)\right)^2} + \frac{Ex+F}{\left(h(x)\right)^3}$
$R(x) =$ $= \left[\left(h(x^2)\right)^3\right] \cdot \left[\frac{Ax+B}{h(x^2)} + \frac{Cx+D}{\left(h(x)\right)^2} + \frac{Ex+F}{\left(h(x)\right)^3}\right]$
$R(x) = (Ax+B) \cdot \left(h(x^2)\right)^2 +$ $+ (Cx+D) \cdot h(x^2) + (Ex+F)$
Solve for A, B, C, D, E, F

Partial Fractions: Case 4 – Ex. 1

Write the partial fraction decomposition form:

$$R(x) = \frac{x^3 + x + 1}{x(x-1)(x^2 + x + 1)(x^2 + 1)^3}$$

$$R(x) = \frac{x^3 + x + 1}{x(x-1)(x^2 + x + 1)(x^2 + 1)^3}$$

$$R(x) = \frac{A}{x} + \frac{B}{(x-1)} + \frac{Cx + D}{(x^2 + x + 1)}$$

$$+ \frac{Ex + F}{(x^2 + 1)^1} + \frac{Gx + H}{(x^2 + 1)^2} + \frac{Ix + J}{(x^2 + 1)^3}$$

$$x^3 + x + 1 =$$

$$A(x - 1)(x^2 + x + 1)(x^2 + 1)^3$$

$$+ Bx(x^2 + x + 1)(x^2 + 1)^3$$

$$+ (Cx + D)x(x - 1)(x^2 + 1)^3$$

$$+ (Ex + F)x(x - 1)(x^2 + x + 1)(x^2 + 1)^2$$

$$+ (Gx + H)x(x - 1)(x^2 + x + 1)(x^2 + 1)^1$$

$$+ (Ix + J)x(x - 1)(x^2 + x + 1)$$

Partial Fractions: Case 4 – Ex. 2a

Evaluate: $\quad I = \int \dfrac{1 - x + 2x^2 - x^3}{x(x^2+1)^2}\, dx$

$$\dfrac{1 - x + 2x^2 - x^3}{x(x^2+1)^2} = \dfrac{A}{x} + \dfrac{Bx+C}{(x^2+1)} + \dfrac{Dx+E}{(x^2+1)^2}$$

$$1 - x + 2x^2 - x^3 = A(x^2+1)^2$$
$$+ (Bx+C)x(x^2+1) + (Dx+E)x$$

$$1 - x + 2x^2 - x^3 = A(x^4+2x^2+1)$$
$$+ (Bx+C)(x^3+x) + (Dx+E)x$$

$1 - x + 2x^2 - x^3 =$ $+\ x^4(A+B)$ $+\ x^3(C)$ $+\ x^2(2A+B+D)$ $+\ x^1(C+E)$ $+\ x^0(A)$	Regroup terms by coefficients of: x^4, x^3, x^2, x^1, x^0 This makes solving for A, B, C, D, and E much easier.

Continued ...

(Stewart, Calculus Early Transcendentals, p. 499)

Partial Fractions: Case 4 – Ex. 2b

Previously, we found...

$1 - x + 2x^2 - x^3 =$

$\quad + x^4(A + B)$

$\quad + x^3(C)$

$\quad + x^2(2A + B + D)$

$\quad + x^1(C + E)$

$\quad + x^0(A)$

x^0	$A = 1$
x^1	$C + E = -1$
x^2	$2A + B + D = 2$
x^3	$C = -1$
x^4	$A + B = 0$

$A = 1,\ B = -1,\ C = -1,\ E = 0,\ D = 1$

$$I = \int \left[\frac{A}{x} + \frac{Bx + C}{(x^2 + 1)} + \frac{Dx + E}{(x^2 + 1)^2} \right] dx$$

$$I = \int \left[\frac{1}{x} - \frac{x + 1}{(x^2 + 1)} + \frac{x}{(x^2 + 1)^2} \right] dx$$

Continued ...

Partial Fractions: Case 4 – Ex. 2c	
Previously, we found...	

$$I = \int \left[\frac{1}{x} - \frac{x+1}{(x^2+1)} + \frac{x}{(x^2+1)^2} \right] dx$$

$$I = \int \left[\frac{1}{x} - \frac{x}{(x^2+1)} - \frac{1}{(x^2+1)} + \frac{x}{(x^2+1)^2} \right] dx$$

$\int \frac{1}{x}\, dx$	$= \ln|x|$
$\int \frac{x}{(x^2+1)}\, dx$	$= \frac{1}{2} \int \frac{1}{u}\, du \qquad \boxed{\begin{array}{l} u = x^2+1 \\ du = 2x\, dx \end{array}}$ $= \frac{1}{2}\ln|u|$ $= \frac{1}{2}\ln(x^2+1)$
$\int \frac{1}{(x^2+1)}\, dx$	$= \tan^{-1}(x)$
$\int \frac{x}{(x^2+1)^2}\, dx$	$= \frac{1}{2} \int u^{-2}\, du \qquad \boxed{\begin{array}{l} u = x^2+1 \\ du = 2x\, dx \end{array}}$ $= \frac{1}{2}\left(\frac{u^{-1}}{-1}\right)$ $= -\frac{1}{2}\left(\frac{1}{x^2+1}\right)$

$$I = \ln|x| - \frac{1}{2}\ln(x^2+1)$$
$$- \tan^{-1}(x) - \frac{1}{2}\left(\frac{1}{x^2+1}\right) + C$$

Partial Fractions: Case 4 – Ex. 3a

Evaluate: $I = \int \frac{\sqrt{x+9}}{x}\, dx$

| This is a non-rational function. Try to change it into a rational function with u-substitution. | $u = \sqrt{x+9}$
 $u^2 = x + 9$
 $x = u^2 - 9$
 $\frac{dx}{du} = 2u$
 $dx = 2u\, du$ |

$I = \int \frac{\sqrt{x+9}}{x}\, dx\ = \int \frac{u}{u^2-9}\, 2u\, du$

$I = 2 \int \frac{u^2}{u^2-9}\, du$

| Long Division | $\ \ 1$
 $u^2 + 0u - 9\, \big|\ \overline{\ u^2 + 0u + 0\ }$
 $\ -\,(u^2 + 0u - 9)$
 $\ \overline{\ \ 9\ }$ |

$I\ =\ 2 \int \left[\frac{u^2}{u^2-9} \right] du\ =\ 2 \int \left[1 + \frac{9}{u^2-9} \right] du$

Continued ...

82

Partial Fractions: Case 4 – Ex. 3b

Evaluate: $I = 2\int \left[1 + \dfrac{9}{u^2 - 9}\right] du$

$$\dfrac{9}{u^2 - 9} = \dfrac{9}{(u+3)(u-3)} = \dfrac{A}{(u+3)} + \dfrac{B}{(u-3)}$$

$9 = A(u - 3) + B(u + 3)$

$u = 3$	$9 = B(6)$	$\rightarrow B = \dfrac{9}{6} = \dfrac{3}{2}$
$u = -3$	$9 = A(-6)$	$\rightarrow A = -\dfrac{9}{6} = -\dfrac{3}{2}$

$I = 2\int \left[1 + \dfrac{A}{(u+3)} + \dfrac{B}{(u-3)}\right] du$

$I = 2\left[u - \left(\dfrac{3}{2}\right)\ln|u+3| + \left(\dfrac{3}{2}\right)\ln|u-3|\right]$

$I = 2u - 3\ln|u+3| + 3\ln|u-3|$

$I = 2u + 3(\ln|u-3| - \ln|u+3|)$

$I = 2u + 3\ln\left|\dfrac{u-3}{u+3}\right|$ $\boxed{u = \sqrt{x+9}}$

$I = 2\sqrt{x+9} + 3\ln\left|\dfrac{\sqrt{x+9}-3}{\sqrt{x+9}+3}\right| + C$

Improper Integrals

Improper Integrals -- Overview

Improper Integrals are definite integrals (boundaries specified) where one or both of the boundaries approach infinity.

Type I	One or both of the boundaries approaches infinity.
Type II	One or both of the boundaries approaches a discontinuous value.

Improper integrals will converge or diverge.

Converge	If the integral converges (settles down) it can be evaluated.
Diverge	If the integral diverges (goes wild) it cannot be evaluated.

Type 1: Infinite Integrals

Improper Integrals – Type 1 (Infinite)	
$\int_a^\infty f(x)\,dx$	$= \lim\limits_{t\to\infty} \int_a^t f(x)\,dx$
$\int_{-\infty}^b f(x)\,dx$	$= \lim\limits_{t\to-\infty} \int_t^b f(x)\,dx$
$\int_{-\infty}^\infty f(x)\,dx$	$= \lim\limits_{t\to-\infty} \int_t^a f(x)\,dx$ $+ \lim\limits_{t\to\infty} \int_a^t f(x)\,dx$
If the limit exists, the integral converges. Otherwise, the integral diverges.	

Improper Integrals: Type 1 – Ex. 1	
Evaluate: $I = \int_1^\infty \frac{1}{x} dx$	
Rewrite integral as a limit.	$I = \lim_{t \to \infty} \int_1^t \frac{1}{x} dx$
Evaluate the integral.	$I = \lim_{t \to \infty} [\ln\|x\|]_1^t$ $I = \lim_{t \to \infty} [\ln\|t\| - \ln\|1\|]$ $I = \ln(\infty) - \ln(1)$ $I = \infty - 0$ $I = \infty \qquad$ DNE
Conclusion	The limit does not exist. The integral is divergent.

88

Improper Integrals: Type 1 – RULE	
$$I = \int_1^\infty \frac{1}{x^p}\, dx$$	
$p > 1$	Convergent
$p \leq 1$	Divergent

Improper Integrals: Type 1 – Ex. 2	
Evaluate: $I = \int_1^\infty \frac{1}{x^2} dx$	
Rewrite integral as a limit.	$I = \lim_{t \to \infty} \int_1^t \frac{1}{x^2} \, dx$ $I = \lim_{t \to \infty} \int_1^t x^{-2} \, dx$
Evaluate the integral.	$I = \lim_{t \to \infty} \left[\frac{x^{-1}}{-1} \right]_1^t$ $I = \lim_{t \to \infty} \left[-\frac{1}{x} \right]_1^t$ $I = \lim_{t \to \infty} \left[-\frac{1}{t} - \left(-\frac{1}{1} \right) \right]$ $I = \lim_{t \to \infty} \left[-\frac{1}{t} + 1 \right]$ $I = \left[-\frac{1}{\infty} + 1 \right] = 0 + 1 = 1$
Conclusion	The limit exists. The integral is convergent. The integral converges to 1.

Improper Integrals: Type 1 – Ex. 3a

Evaluate: $I = \int_1^\infty x e^{-x}\, dx$

Rewrite integral as a limit.	$I = \lim_{t \to \infty} \int_1^t x e^{-x}\, dx$

Evaluate the integral. Use integration by parts (tabular method)

$\frac{d}{dx}$		$\int$
x	$+$	e^{-x}
1	$-$	$-e^{-x}$
0	$+$	e^{-x}

$I = \lim_{t \to \infty} [-x e^{-x} - e^{-x}]_1^t$

$I = \lim_{t \to \infty} [-e^{-x}(x+1)]_1^t$

$I = \lim_{t \to \infty} \left[-\frac{x+1}{e^x}\right]_1^t$

$I = \lim_{t \to \infty} \left[\left(-\frac{t+1}{e^t}\right) - \left(-\frac{1+1}{e^1}\right)\right]$

$I = \frac{\infty}{\infty} + \frac{2}{e}$

Limit of first term is indeterminate so use L'Hopital's Rule.   Continued …

Improper Integrals: Type 1 – Ex. 3b
Evaluate: $I = \int_1^\infty xe^{-x}\,dx$

$I = \lim\limits_{t \to \infty} \left[\left(-\frac{t+1}{e^t}\right) - \left(-\frac{1+1}{e^1}\right)\right]$

$I = \frac{\infty}{\infty} + \frac{2}{e}$

$I = \lim\limits_{t \to \infty} \left[\left(-\frac{1+0}{e^t}\right) + \left(\frac{2}{e}\right)\right]$

$I = \left(-\frac{1}{\infty}\right) + \left(\frac{2}{e}\right)$

$I = 0 + \frac{2}{e} = \frac{2}{e}$

LH Rule
Take deriv. of numerator and denom.

Conclusion	The limit exists.
	The integral is convergent.
	The integral converges to $\frac{2}{e}$

Improper Integrals: Type 1 – Ex. 4
Evaluate: $I = \int_{-\infty}^{\infty} \frac{1}{9+x^2}\, dx$

$I = \lim\limits_{t \to -\infty} \int_{t}^{0} \frac{1}{9+x^2}\, dx \; + \; \lim\limits_{t \to \infty} \int_{0}^{t} \frac{1}{9+x^2}\, dx$

$$\int \frac{du}{a^2+u^2} = \frac{1}{a}\tan^{-1}\left(\frac{u}{a}\right)$$

$I = \lim\limits_{t \to -\infty} \left[\frac{1}{3}\tan^{-1}\left(\frac{x}{3}\right)\right]_{t}^{0} + \lim\limits_{t \to \infty} \left[\frac{1}{3}\tan^{-1}\left(\frac{x}{3}\right)\right]_{0}^{t}$

$I = \frac{1}{3}[\,0 - \tan^{-1}(-\infty)\,] + \frac{1}{3}[\,\tan^{-1}(\infty) - 0\,]$

$I = \frac{1}{3}\left[0 - \left(-\frac{\pi}{2}\right)\right] + \frac{1}{3}\left[\frac{\pi}{2} - 0\right]$

$I = \frac{\pi}{6} + \frac{\pi}{6} = \frac{2\pi}{6} = \frac{\pi}{3}$

Conclusion	The limit exists.
	The integral is convergent.
	The integral converges to $\frac{\pi}{3}$

<u>Type 2: Discontinuous Integrals</u>

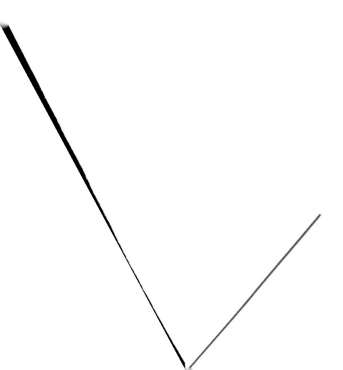

Improper Integrals – Type 2 (Discontinuous)	
$\displaystyle\int_a^b f(x)\, dx$ Discontinuous at b	$= \displaystyle\lim_{t \to b^-} \int_a^t f(x)\, dx$
$\displaystyle\int_a^b f(x)\, dx$ Discontinuous at a	$= \displaystyle\lim_{t \to a^+} \int_t^b f(x)\, dx$
$\displaystyle\int_a^b f(x)\, dx$ Discontinuous at c Where: $a < c < b$	$= \displaystyle\lim_{t \to a^-} \int_t^c f(x)\, dx$ $+ \displaystyle\lim_{t \to b} \int_c^t f(x)\, dx$
If the limit exists, the integral converges. Otherwise, the integral diverges.	

Improper Integrals: Type 2 – Ex. 1	
Evaluate: $I = \int_3^7 \frac{1}{\sqrt{x-3}}\,dx$	
Rewrite as a limit.	$I = \lim_{t \to 3^-} \int_t^7 \frac{1}{\sqrt{x-3}}\,dx$
Evaluate the integral.	$I = \lim_{t \to 3^-} \int_t^7 (x-3)^{-\frac{1}{2}}\,dx$ $I = \lim_{t \to 3^-} \left[\frac{(x-3)^{\frac{1}{2}}}{\left(\frac{1}{2}\right)} \right]_t^7$ $I = \lim_{t \to 3^-} 2\left[\sqrt{x-3} \right]_t^7$ $I = \lim_{t \to 3^-} 2\left[\sqrt{4} - \sqrt{3-t} \right]$ $I = 2[\,2-0\,] = 4$
Conclusion	The limit exists. Integral is convergent. Integral converges to 4.

Improper Integrals: Type 2 – Ex. 2	
Evaluate: $I = \int_0^5 \dfrac{1}{x-3}\,dx$	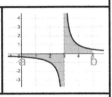
$I = \lim\limits_{t \to 3} \int_0^t \dfrac{1}{x-3}\,dx \;+\; \lim\limits_{t \to 3^-} \int_t^5 \dfrac{1}{x-3}\,dx$	
$I = \lim\limits_{t \to 3} [\,\ln|x-3|\,]_0^t \;+\; \lim\limits_{t \to 3^-} [\,\ln|x-3|\,]_t^5$ $I = [\,\ln(0) - \ln|-3|\,] \;+\; [\,\ln(2) - \ln(0)\,]$ $I = [-\infty - \ln(3)\,] \;+\; [\,\ln(2) - \infty\,]$ $I = \;-\infty \;-\; \infty \;=\; -\infty$	
Conclusion	The limit does not exist. Integral is divergent.

Improper Integrals: Type 2 – Ex. 3a	
Evaluate: $I = \int_0^1 \ln x\, dx$	

Rewrite as a limit.	$I = \lim\limits_{t \to 0^+} \int_t^1 \ln x\ dx$
Integrate by Parts: $u = \ln x$ $dv = dx$	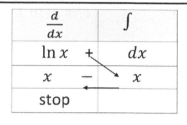 $I = \int \ln x\, dx$ $I = (\ln x)(x) - x$

$I = \lim\limits_{t \to 0^+} \int_t^1 \ln x\ dx$

$I = \lim\limits_{t \to 0^+} [\,x \ln x - x\,]_t^1$

Continued ...

Improper Integrals: Type 2 – Ex. 3b	
Evaluate: $I = \int_0^1 \ln x \, dx$	

From previous page	$I = \lim_{t \to 0^+} [\, x \ln x - x \,]_t^1$
	$I = [(\ln 1 - 1) - (0)]$
	$I = 0 - 1 = -1$
Conclusion	The limit exists.
	The integral converges.
	Integral converges to -1

Comparison Test for Improper Integrals

Comparison Theorem
Given: $f(x) \geq g(x) \geq 0$ for $x \geq a$
If: $\int_a^\infty f(x)\,dx$ is convergent,   Then: $\int_a^\infty g(x)\,dx$ is convergent.
If: $\int_a^\infty g(x)\,dx$ is divergent,   Then: $\int_a^\infty f(x)\,dx$ is divergent.
Note: It's just common sense!

Improper Integrals: Comparison Test – Ex. 1
$I = \int_1^\infty e^{-x^2}\, dx$ Does it converge?
Find a similar and larger function $f_2(x)$, in the interval, that converges.

$I_1 = \int_1^\infty e^{-x^2}\, dx$

$I_2 = \int_1^\infty e^{-x}\, dx$

$f_1(x) = e^{-x^2}$

$f_2(x) = e^{-x}$

$I_2 = \int_1^\infty e^{-x}\, dx = \lim_{t \to \infty} \int_1^t e^{-x}\, dx$

$I_2 = \lim_{t \to \infty} \left[-e^{-x} \right]_1^t = \lim_{t \to \infty} \left[-\frac{1}{e^x} \right]_1^t$

$I_2 = \lim_{t \to \infty} \left[\left(-\frac{1}{e^t} \right) - \left(-\frac{1}{e} \right) \right] = \left[0 + \frac{1}{e} \right] = \frac{1}{e}$

Conclusion	I_2 Converges. $I_1 \leq I_2$ on the interval. I_1 Converges by C.T.

Improper Integrals: Comparison Test – Ex. 2
$I = \int_1^\infty \dfrac{1 + e^{-x^2}}{x} dx$ Does it diverge?
Find a similar and smaller larger function $f_2(x)$, in the interval, that diverges.
$f_1 = \dfrac{1 + e^{-x^2}}{x} = \dfrac{1}{x} + \dfrac{e^{-x^2}}{x}$ $f_2 = \dfrac{1}{x} \quad \rightarrow \quad f_2 \leq f_1$ in the interval
Let: $I = I_1 = \int_1^\infty \dfrac{1 + e^{-x^2}}{x} dx$
$I_2 = \int_1^\infty \dfrac{1}{x} dx = \lim\limits_{t \to \infty} \int_1^t e^{-x} dx$ $I_2 = \lim\limits_{t \to \infty} [\ln x]_1^t = \lim\limits_{t \to \infty}[\ln\lvert t\rvert - \ln(1)]$ $I_2 = [(\infty) - (0)] = \infty$
Conclusion I_2 Diverges. $I_1 \geq I_2$ on the interval. I_1 Diverges by C.T.

More Applications of Integration

Arc Length

Arc Length -- Formulas	
For $y = f(x)$	$L = \int_{x=a}^{x=b} \sqrt{1 + [f'(x)]^2}\, dx$
For $x = f(y)$	$L = \int_{y=a}^{y=b} \sqrt{1 + [f'(y)]^2}\, dy$
Note: Formulas are based on the distance formula or The Pythagorean Theorem: $$c^2 = a^2 + b^2$$ $$c = \sqrt{a^2 + b^2}$$	

Arc Length – Ex. 1a

Find the length of the arc on the curve $y = x^2$ between: $(1,1)$ and $(3,9)$	

$y = f(x) = x^2$ $y' = f'(x) = 2x$	$L = \int_{x=a}^{x=b} \sqrt{1 + [f'(x)]^2} \, dx$ $L = \int_{x=1}^{x=3} \sqrt{1 + (2x)^2} \, dx$
Use trig substitution Here: $a = 1$ $\quad\quad u = 2x$	$\sqrt{u^2 + a^2}$ $\quad u$ θ a
Trig Substitutions	$\cos \theta = \dfrac{1}{\sqrt{1 + 4x^2}}$ $\sqrt{1 + 4x^2} = \dfrac{1}{\cos \theta} = \sec \theta$ $\tan \theta = \dfrac{2x}{1} = 2x$ $x = \dfrac{1}{2}\tan \theta$ $\dfrac{dx}{d\theta} = \dfrac{1}{2}\sec^2 \theta$ $dx = \dfrac{1}{2}\sec^2 \theta \, d\theta$
Continued ...	

Arc Length – Ex. 1b

Find the length of the arc
on the curve $y = x^2$
between: $(1, 1)$ and $(3, 9)$

$y = x^2$ $y' = 2x$	$L = \int_{x=1}^{x=3} \sqrt{1 + (2x)^2} \; dx$
Use trig substitutions from previous page.	$L = \int (\sec \theta) \left(\frac{1}{2} \sec^2 \theta \; d\theta \right)$ $L = \frac{1}{2} \int \sec^3 \theta \; d\theta$ Use Formula

$$\int \sec^3 u \; du = \frac{1}{2} [\sec u \cdot \tan u + \ln| \sec u + \tan u |]$$

$$L = \frac{1}{4} [\sec \theta \tan \theta + \ln|\sec \theta + \tan \theta|]$$

Use diagram to convert.
$a = 1, \; u = 2x$

$\sec \theta = \sqrt{1 + 4x^2}$
$\tan \theta = 2x$

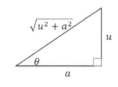

$$L = \frac{1}{4} \left[2x\sqrt{1 + 4x^2} + \ln\left|\sqrt{1 + 4x^2} + 2x\right| \right]$$

Continued ...

Arc Length – Ex. 1c

Find the length of the arc on the curve $y = x^2$ between: $(1, 1)$ and $(3, 9)$

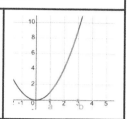

Previously found:

$$L = \frac{1}{4}\left[2x\sqrt{1 + 4x^2} + \ln\left|\sqrt{1 + 4x^2} + 2x\right|\right]$$

Evaluate between $x = 1$ and $x = 3$

$x = 1$	$L_1 = \frac{1}{4}\left[2\sqrt{1 + 4} + \ln\left|\sqrt{1 + 4} + 2\right|\right]$
	$L_1 = \frac{1}{4}\left[2\sqrt{5} + \ln\left|\sqrt{5} + 2\right|\right]$
	$L_1 \approx \frac{1}{4}[5.92] \approx 1.48$

$x = 3$	$L_3 = \frac{1}{4}\left[6\sqrt{37} + \ln\left|\sqrt{37} + 6\right|\right]$
	$L_3 \approx \frac{1}{4}[38.99] \approx 9.75$

$L = L_3 - L_1$

$L \approx 9.75 - 1.48 \approx 8.27$

Arc Length – Ex. 2

Find the length of the arc on the curve

$$y = 1 + 6x^{\frac{3}{2}} \ ; \ \ 0 \leq x \leq 1$$

$y = 1 + 6x^{\frac{3}{2}}$ $y' = 9x^{\frac{1}{2}}$	$L = \int_{x=a}^{x=b} \sqrt{1 + [\,f'(x)]^2}\ dx$ $L = \int_{x=0}^{x=1} \sqrt{1 + \left(9\sqrt{x}\,\right)^2}\ dx$ $L = \int_0^1 \sqrt{1 + 81x}\ dx$
Use u-sub $u = 1 + 81x$ $du = 81\ dx$	$L = \left(\frac{1}{81}\right) \int u^{\frac{1}{2}}\ du$ $L = \left(\frac{1}{81}\right)\left[\left(\frac{2}{3}\right) u^{\frac{3}{2}}\right]$ $L = \left(\frac{2}{243}\right)\left[u^{\frac{3}{2}}\right]$
Substitute Back.	$L = \left(\frac{2}{243}\right)\left[(1 + 81x)^{\frac{3}{2}}\right]_0^1$ $L = \left(\frac{2}{243}\right)\left[(82)^{\frac{3}{2}} - (1)^{\frac{3}{2}}\right]$ $L = \left(\frac{2}{243}\right)\left[82\sqrt{82} - 1\right]$ $L \approx \left(\frac{2}{243}\right)[741.54] \approx 6.10$

Arc Length – Ex. 3

Find equation for length of the arc on the curve.
Then, evaluate the integral with a calculator.
$$x = y^2 - 2y \;\; ; \;\; 0 \le y \le 2$$

$x = y^2 - 2y$

$x' = 2y - 2 = 2(y - 1)$

$(x')^2 = 4(y - 1)^2 = 4(y^2 - 2y + 1)$

$(x')^2 = 4y^2 - 8y + 4$

$L = \int_0^2 \sqrt{1 + (x')^2} \; dy$

$L = \int_0^2 \sqrt{4y^2 - 8y + 5} \; dy$

Evaluated with
DESMOS online
graphing calculator

$L \approx 2.96$

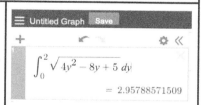

Note:
To complete this problem without a calculator,
complete the square to get the form:
$\sqrt{(u)^2 + a}$ then use trig substitutions.

Area of Surface of Revolution

Surface Area of Revolution -- Formulas
Circumference $= 2\pi r$ Surface Area $= 2\pi r l$
$S = $ Surface Area $S = $ (Circumference)·(Length) $S = 2\pi r \cdot L \;\; = \int_a^b (2\pi r) \cdot L$
For: $y = f(x)$ 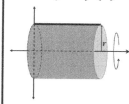 $S = \int_a^b 2\pi y \sqrt{1 + [y']^2}\; dx$
For: $x = f(y)$ 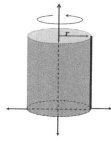 $S = \int_a^b 2\pi x \sqrt{1 + [x']^2}\; dy$

Surface Area of Revolution – Ex. 1

Find the area of the surface
obtained by rotating this arc
about the x-axis.

$y = \sqrt{3 - x^2}, \ -1 \le x \le 1$

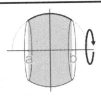

First, get the parts needed for the equation.

$y = (3 - x^2)^{\frac{1}{2}}$

$y' = \frac{1}{2}(3 - x^2)^{-\frac{1}{2}}(-2x) = \frac{-x}{\sqrt{3 - x^2}}$

$(y')^2 = \frac{x^2}{(3 - x^2)}$

$1 + (y')^2 = \frac{(3 - x^2)}{(3 - x^2)} + \frac{x^2}{(3 - x^2)} = \frac{3}{(3 - x^2)}$

$S = \int_a^b 2\pi y \cdot \sqrt{1 + [y']^2}\ dx$ Use this eqn.

$S = \int_{-1}^{1} 2\pi \sqrt{3 - x^2} \cdot \sqrt{\frac{3}{(3 - x^2)}}\ dx$

$S = 2\pi \int_{-1}^{1} \sqrt{3 - x^2} \cdot \frac{\sqrt{3}}{\sqrt{3 - x^2}}\ dx$

$S = 2\pi\sqrt{3} \cdot \int_{-1}^{1} 1\ dx$

$S = 2\pi\sqrt{3}\,[\,x\,]_{-1}^{1} = 2\pi\sqrt{3}\,[(1) - (-1)]$

$S = 2\pi\sqrt{3}\,[\,2\,] = 4\pi\sqrt{3}$

113

Surface Area of Revolution – Ex. 2a

Find the area of the surface
obtained by rotating this arc
about the y-axis.

$$y = x^2 , \quad 1 \le x \le 3$$

$S = \int_a^b 2\pi x \cdot \sqrt{1 + [\,x'\,]^2}\ dy$ Use this eqn.

First, get the parts needed for the equation.

$$x = (y)^{\frac{1}{2}}$$

$$x' = \frac{1}{2}(y)^{-\frac{1}{2}} = \frac{1}{2\sqrt{y}}$$

$$(x')^2 = \frac{1}{4y}$$

$$1 + (x')^2 = \frac{4y}{4y} + \frac{1}{4y} = \frac{4y+1}{4y}$$

$$S = \int_{y=a}^{y=b} 2\pi x \cdot \sqrt{1 + [\,x'\,]^2}\ dy$$

$$S = \int_{y=1}^{y=9} 2\pi \sqrt{y} \cdot \sqrt{\frac{4y+1}{4y}}\ dy \qquad \begin{array}{l} x = 1 \\ y = 1^2 = 1 \end{array}$$

$$S = 2\pi \int_1^9 \sqrt{y} \cdot \frac{\sqrt{4y+1}}{2\sqrt{y}}\ dy \qquad \begin{array}{l} x = 3 \\ y = 3^2 = 9 \end{array}$$

$$S = \pi \cdot \int_1^9 \sqrt{4y + 1}\ dy$$

Continued …

114

Surface Area of Revolution – Ex. 2b

Find the area of the surface obtained by rotating this arc about the y-axis.

$$y = x^2 \ , \ \ 1 \leq x \leq 3$$

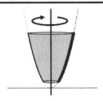

$S = \pi \cdot \int_1^9 \sqrt{4y + 1} \ dy$ Found previously

Use u-sub	$u = 4y + 1$
	$\frac{du}{dy} = 4 \quad \rightarrow \quad du = 4 \, dy$

$S = \pi \cdot \int_1^9 \sqrt{4y + 1} \ dy$

$S = \frac{\pi}{4} \cdot \int \sqrt{u} \ du$

$S = \frac{\pi}{4} \left[\left(\frac{2}{3} \right) u^{\frac{3}{2}} \right] = \frac{\pi}{6} \left[u^{\frac{3}{2}} \right]$

$S = \frac{\pi}{6} \left[(4y + 1)^{\frac{3}{2}} \right]_1^9$

$S = \frac{\pi}{6} \left[(37)^{\frac{3}{2}} - (5)^{\frac{3}{2}} \right] \approx \frac{\pi}{6} [213.88]$

$S \approx 111.99$

Surface Area of Revolution – Ex. 3

Find the area of the surface obtained by rotating this arc about the x-axis.

$$y = x^3 , \ 0 \leq x \leq 2$$

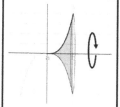

First, get the parts needed for the equation.

$$S = \int_a^b 2\pi y \cdot \sqrt{1 + [\,y'\,]^2} \ dx \qquad \text{Use this eqn.}$$

$$y = x^3$$

$$y' = 3x^2$$

$$(y')^2 = (3x^2)^2 = 9x^4$$

$$1 + (y')^2 = 1 + (3x^2)^2 = 1 + 9x^4$$

$$S = \int_0^2 2\pi\, x^3 \cdot \sqrt{1 + 9x^4}\ dx \qquad \boxed{\begin{aligned} u &= 1 + 9x^4 \\ du &= 36\,x^3 dx \end{aligned}}$$

$$S = 2\pi \int_0^2 x^3\sqrt{1 + 9x^4}\ dx$$

$$S = \frac{2\pi}{36} \int u^{\frac{1}{2}} \ du \ = \ \frac{\pi}{18}\left[\left(\tfrac{2}{3}\right)u^{\frac{3}{2}}\right]$$

$$S = \frac{\pi}{27}\left[u^{\frac{3}{2}}\right] \ = \ \frac{\pi}{27}\left[(1+9x^4)^{\frac{3}{2}}\right]_0^2$$

$$S = \frac{\pi}{27}\left[(145)^{\frac{3}{2}} - (1)\right] \approx 203.04$$

Center of Mass

Center of Mass -- Equations	
A plate with uniform density has a center of mass, or centroid, at the point $(\bar{x}, \bar{y})$. $A =$ Area of the region	
Region is defined by $y = f(x)$	$\bar{x} = \frac{1}{A} \int_a^b x \cdot f(x)\, dx$ $\bar{y} = \frac{1}{A} \int_a^b \frac{1}{2} [f(x)]^2\, dx$
Region lies between 2 curves: $y_1 = f(x)$ $y_2 = g(x)$	$\bar{x} = \frac{1}{A} \int_a^b x\,[y_1 - y_2]\, dx$ $\bar{y} = \frac{1}{A} \int_a^b \frac{1}{2} [\,(y_1)^2 - (y_2)^2\,]\, dx$

Center of Mass -- Ex. 1a

Find the center of mass of a semicircular plate with a radius of 5.	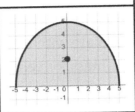

Area	$A = \frac{1}{2} [\pi r^2]$ $A = \frac{1}{2} [\pi 5^2] = \frac{25\pi}{2}$
$y = f(x)$	$x^2 + y^2 = r^2$ Circle $y = \pm\sqrt{r^2 - x^2}$ $y = \sqrt{5^2 - x^2}$ Semi-Circle
$\bar{x}$	$\bar{x} = 0$ By symmetry
$\bar{y}$	$\bar{y} = \frac{1}{A} \int_a^b \frac{1}{2} [f(x)]^2 \, dx$ $\bar{y} = \frac{2}{25\pi} \int_{-5}^{5} \frac{1}{2} [5^2 - x^2] \, dx$ $\bar{y} = \frac{1}{25\pi} \int_{-5}^{5} [5^2 - x^2] \, dx$
	Continued ...

Center of Mass -- Ex. 1b	
Find the center of mass of a semicircular plate with a radius of 5.	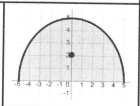

... Continued ...	
$\bar{y}$	$\bar{y} = \frac{1}{25\pi} \int_{-5}^{5} [5^2 - x^2]\, dx$ $\bar{y} = \frac{2}{25\pi} \int_{0}^{5} [5^2 - x^2]\, dx$ $\bar{y} = \frac{2}{25\pi} \left[25x - \frac{x^3}{3} \right]_0^5$ $\bar{y} = \frac{2}{25\pi} \left[125 - \frac{125}{3} \right]$ $\bar{y} = \frac{2(125)}{25\pi} \left[1 - \frac{1}{3} \right]$ $\bar{y} = \frac{2(5)}{\pi} \left[\frac{2}{3} \right]$ $\bar{y} = \frac{20}{3\pi} \approx 2.1$
$(\bar{x}, \bar{y})$	$(\bar{x}, \bar{y}) = \left(0, \frac{20}{3\pi} \right) \approx (0, 2.1)$

Center of Mass -- Ex. 2a

Find the centroid of the region bounded by the curves:
$y = \sin x$, $y = 0$, and
$x = 0$, $x = \dfrac{\pi}{2}$

Area	$A = \int_0^{\frac{\pi}{2}} \sin x\, dx = [-\cos x]_0^{\frac{\pi}{2}}$ $A = -\left[\cos\dfrac{\pi}{2} - \cos 0\right] = 1$

$\bar{x}$

$$\bar{x} = \frac{1}{A}\int_a^b x \cdot f(x)\, dx$$

$$\bar{x} = \int_0^{\frac{\pi}{2}} x \cdot \sin x\, dx$$

dx	$\int$
x +	$\sin x$
1 −	$-\cos x$
0 +	$-\sin x$

$$\bar{x} = [-x\cos x + \sin x]_0^{\frac{\pi}{2}}$$

$$\bar{x} = \sin\frac{\pi}{2} = 1$$

Continued ...

121

Center of Mass -- Ex. 2b	
Find the centroid of the region bounded by the curves: $y = \sin x$, $y = 0$, and $x = 0$, $x = \dfrac{\pi}{2}$	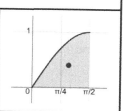

... Continued ...	
$\bar{y}$	$\bar{y} = \dfrac{1}{A}\int_a^b \dfrac{1}{2}\,[\,f(x)\,]^2\ dx$ $\bar{y} = \int_0^{\frac{\pi}{2}} \dfrac{1}{2}\,[\ \sin x\ ]^2\ dx$ $\bar{y} = \dfrac{1}{2}\int_0^{\frac{\pi}{2}}\left[\dfrac{1}{2}(1 - \cos 2x\,)\right]\ dx$ $\bar{y} = \dfrac{1}{4}\int_0^{\frac{\pi}{2}} (1 - \cos 2x\,)\ dx$ $\bar{y} = \dfrac{1}{4}\left[\ x - \dfrac{1}{2}\sin 2x\ \right]_0^{\frac{\pi}{2}}$ $\bar{y} = \dfrac{1}{4}\left[\dfrac{\pi}{2} - \dfrac{1}{2}\sin \pi\ \right]$ $\bar{y} = \dfrac{1}{8}\,[\ \pi\ -\ 0\,]\ =\ \dfrac{\pi}{8}\ \approx\ 0.4$
$(\bar{x}, \bar{y})$	$(\bar{x}, \bar{y}) = \left(1,\ \dfrac{\pi}{8}\right)\ \approx\ (1, 0.4\,)$

Center of Mass -- Ex. 3a

Find the centroid of the region bounded by the curves:

$y_1 = 2x$ and $y_2 = x^2$

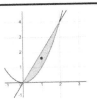

Inter-section	$y_1 = y_2$ $2x = x^2$ $2x - x^2 = 0$ $x(2-x) = 0 \quad \rightarrow \quad x = 0, 2$
Area	$A = \int_0^2 (y_1 - y_2)\, dx$ $A = \int_0^2 (2x - x^2)\, dx$ $A = \left[x^2 - \dfrac{x^3}{3} \right]_0^2$ $A = \left[4 - \dfrac{8}{3} \right] = \left[\dfrac{12}{3} - \dfrac{8}{3} \right]$ $A = \dfrac{4}{3}$

Continued ...

123

Center of Mass -- Ex. 3b

Find the centroid of the region
bounded by the curves:

$y_1 = 2x$ and $y_2 = x^2$

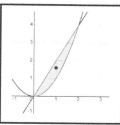

... Continued ...

$\bar{x} = \frac{1}{A} \int_a^b x\,[\,y_1 - y_2\,]\,dx$

$\bar{x} = \frac{3}{4} \int_0^2 x\,[\,2x - x^2\,]\,dx$

$\bar{x} = \frac{3}{4} \int_0^2 [\,2x^2 - x^3\,]\,dx$

$\bar{x} = \frac{3}{4} \left[\,\frac{2}{3}\,x^3 - \frac{1}{4}\,x^4\,\right]_0^2$

$\bar{x} = \frac{3}{4} \left[\,\frac{16}{3} - \frac{16}{4}\,\right]$

$\bar{x} = \frac{3}{4} \left[\,\frac{4}{3}\,\right] = 1$

$\bar{x}$

Continued ...

Center of Mass -- Ex. 3c

Find the centroid of the region bounded by the curves:

$y_1 = 2x$ and $y_2 = x^2$

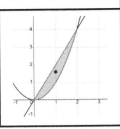

... Continued ...

$\bar{y}$	$\bar{y} = \frac{1}{A} \int_a^b \frac{1}{2} [(y_1)^2 - (y_2)^2]\, dx$
	$\bar{y} = \frac{3}{4} \int_0^2 \frac{1}{2} [4x^2 - x^4]\, dx$
	$\bar{y} = \frac{3}{8} \int_0^2 [4x^2 - x^4]\, dx$
	$\bar{y} = \frac{3}{8} \left[\frac{4}{3}x^3 - \frac{1}{5}x^5 \right]_0^2$
	$\bar{y} = \frac{3}{8} \left[\frac{32}{3} - \frac{32}{5} \right] = \frac{3}{8} \left[\frac{64}{15} \right]$
	$\bar{y} = \frac{1}{8} \left[\frac{64}{5} \right] = \frac{8}{5} \approx 1.6$
$(\bar{x}, \bar{y})$	$(\bar{x}, \bar{y}) = \left(1, \frac{8}{5} \right) \approx (1, 1.6)$

Center of Mass -- Ex. 4a

Find the centroid of the region bounded by the curves:

$y_1 = 1$, $y_2 = 2$, $y_3 = \sqrt{x}$

And the y axis.

Note	Integral should be taken sideways, or $\int f(y)\ dy$ Change y_2 into a function of y Use: $x = f(y) = y^2$
Area	$A = \int_{y=1}^{y=2} (\,y^2\,)\,dy$ $A = \left[\dfrac{y^3}{3}\right]_1^2 = \left[\dfrac{8}{3} - \dfrac{1}{3}\right] = \dfrac{7}{3}$

Continued ...

(Anton, Calculus Early Transcendentals
Single Variable, p. 464)

Center of Mass -- Ex. 4b

Find the centroid of the region bounded by the curves: $y_1 = 1$, $y_2 = 2$, $y_3 = \sqrt{x}$ And the y axis.	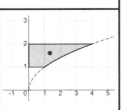

Previously we found this.	$A = \dfrac{7}{3}$ and $x = y^2$

$\bar{x}$ Equations must be revised to integrate sideways	$\bar{y} = \dfrac{1}{A} \int_a^b \dfrac{1}{2} [f(x)]^2 \, dx$
	$\bar{x} = \dfrac{1}{A} \int_{y=a}^{y=b} \dfrac{1}{2} [f(y)]^2 \, dy$
	$\bar{x} = \dfrac{3}{7} \int_{y=1}^{y=2} \dfrac{1}{2} [y^2]^2 \, dy$
	$\bar{x} = \dfrac{3}{14} \left[\dfrac{y^5}{5} \right]_{y=1}^{y=2}$
	$\bar{x} = \dfrac{3}{70} [y^5]_{y=1}^{y=2}$
	$\bar{x} = \dfrac{3}{70} [32 - 1] = \dfrac{93}{70} \approx 1.3$

Continued...

Center of Mass -- Ex. 4c	
Find the centroid of the region bounded by the curves: $y_1 = 1$, $y_2 = 2$, $y_3 = \sqrt{x}$ And the y axis.	

Previously we found this.	$A = \dfrac{7}{3}$ and $x = y^2$ $\bar{x} = \dfrac{93}{70} \approx 1.3$
$\bar{y}$ Equations must be revised to integrate sideways.	$\bar{x} = \dfrac{1}{A} \int_a^b x \cdot f(x)\ dx$ $\bar{y} = \dfrac{1}{A} \int_{y=a}^{y=b} y \cdot f(y)\ dy$ $\bar{y} = \dfrac{3}{7} \int_{y=1}^{y=2} y \cdot y^2\ dy$ $\bar{y} = \dfrac{3}{7} \left[\dfrac{y^4}{4} \right]_{y=1}^{y=2}$ $\bar{y} = \dfrac{3}{28} [16 - 1] = \dfrac{45}{28} \approx 1.6$
$(\bar{x}, \bar{y})$	$(\bar{x}, \bar{y}) = (1.3, 1.6)$

Center of Mass -- Ex. 5a

Find the centroid of the region bounded by the curves:

$y_1 = 2$, $y_2 = \sqrt{x}$

And $x = 1$

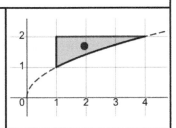

Note	We can get area with either: $\int f(y)\ dy$ or $\int f(x)\ dx$
Area	$A = \int_{x=1}^{x=4} \left(2 - \sqrt{x} \right) dx$ $A = \int_{x=1}^{x=4} \left(2 - x^{\frac{1}{2}} \right) dx$ $A = \left[2x - \left(\frac{2}{3}\right) x^{\frac{3}{2}} \right]_{1}^{4}$ $A = \left(\frac{2}{3}\right) \left[3x - x^{\frac{3}{2}} \right]_{1}^{4}$ $A = \left(\frac{2}{3}\right) \left[(4) - (2) \right]$ $A = \left(\frac{2}{3}\right) [2] = \frac{4}{3}$ $A \approx 1.333$

Continued ...

129

Center of Mass -- Ex. 5b	
Find the centroid of the region bounded by the curves: $y_1 = 2$, $y_2 = \sqrt{x}$ And $x = 1$	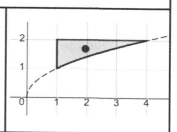
Previously found	$A \approx 1.333$
$\bar{y}$	$\bar{y} = \frac{1}{A} \int_a^b \left(\frac{1}{2}\right) [f(x)]^2 \, dx$

$$\bar{y} = \frac{1}{1.333} \int_{x=1}^{x=4} \left(\frac{1}{2}\right) \left[(2)^2 - (\sqrt{x})^2\right] dx$$

$$\bar{y} = \frac{1}{2.666} \int_1^4 [4 - x] \, dx$$

$$\bar{y} = \frac{1}{2.666} \left[4x - \frac{1}{2}x^2\right]_1^4$$

$$\bar{y} = \frac{1}{2.666} [(8) - (3.5)]$$

$$\bar{y} \approx 1.69$$

Continued...

Center of Mass -- Ex. 5c

Find the centroid of the region bounded by the curves:

$y_1 = 2$, $y_2 = \sqrt{x}$

And $x = 1$

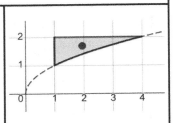

Previously found	$A \approx 1.333 \qquad \bar{y} \approx 1.69$
$\bar{x}$	$\bar{x} = \frac{1}{A} \int_a^b x \cdot f(x)\ dx$

$$\bar{x} = \frac{1}{1.333} \int_{x=1}^{x=4} x \cdot \left[\, 2 - \sqrt{x} \,\right]\ dx$$

$$\bar{x} = \frac{1}{1.333} \int_{x=1}^{x=4} \left[\, 2x - x^{\left(\frac{3}{2}\right)} \,\right]\ dx$$

$$\bar{x} = \frac{1}{1.333} \left[\, x^2 - \left(\frac{2}{5}\right) x^{\left(\frac{5}{2}\right)} \,\right]_1^4$$

$$\bar{x} = \frac{1}{1.333} \left[\, (3.2) - (0.6) \,\right] \quad \approx \quad 1.95$$

$(\bar{x}, \bar{y})$	$(\bar{x}, \bar{y}) \approx (\, 1.95,\ 1.69 \,)$

Center of Mass
Theorem of Pappus

If

R is a bounded plane region

and L is a line that lies in the plane

and R is entirely on one side of L

Then

The volume of the solid formed by revolving R about L is given by:

$$V = A \cdot D$$

Where:

V = Volume

A = Area of R

D = Distance traveled by centroid

Center of Mass
Theorem of Pappus -- Ex. 1

Find the VOLUME of the region bounded by the curves:

$y_1 = 2$, $y_2 = \sqrt{x}$
And $x = 1$

And rotated about the y-axis

Note	In example 5, we found the area and center of mass. $A \approx 1.33$ $(\bar{x}, \bar{y}) \approx (1.95, 1.69)$
Distance Traveled	$D =$ Distance Traveled $D = 2\pi r$ $D = 2\pi\,\bar{x}$ $D \approx 2\pi\,(1.95) = 12.25$
Volume	$V = A \cdot D$ $V \approx (1.33) \cdot (12.25)$ $V \approx 16.34$

Differential Equations (DE)

Differential equations are a powerful and useful tools in modeling and solving many real-world problems. This section includes some DE examples and are just an introduction. For more information read additional books, devoted to the topic.

Naturally, I recommend my own book: "Differential Equations With Applications: Class Notes With Detailed Examples, Patel and Paulk." I wrote this book, based on my class notes, after taking the class.

DE: Separable Equations

Differential Equation Equations (DE) Separable Equations	
$\dfrac{dy}{dx} = g(x) \cdot f(y)$	
How to solve	$\dfrac{dy}{dx} = g(x) \cdot f(y)$ $dy = g(x) \cdot f(y) \cdot dx$ $\dfrac{1}{f(y)} \, dy = g(x) \cdot dx$ $\text{Let: } h(y) = \dfrac{1}{f(y)}$ $h(y) \cdot dy = g(x) \cdot dx$ $\int h(y) \, dy = \int g(x) \, dx$

DE: Separable Equations – Ex. 1

Solve the DE: $\dfrac{dy}{dx} = \dfrac{x^2}{y}$, $y > 0$

Then, find the solution that satisfies
the initial condition, $y(0) = 3$

Rearrange	$\dfrac{dy}{dx} = x^2 \left(\dfrac{1}{y}\right)$ $y \cdot dy = x^2 \cdot dx$
Integrate	$\int y\,dy = \int x^2\,dx$ $\dfrac{1}{2}y^2 = \dfrac{1}{3}x^3 + C_1$ $y^2 = \dfrac{2}{3}x^3 + C$ $y = \sqrt{\dfrac{2}{3}x^3 + C}$, $y > 0$
Solve for "C" Use: $y(0) = 3$	$y = \sqrt{\dfrac{2}{3}x^3 + C}$ $3 = \sqrt{C}$ $C = 9$
Particular Solution	$y = \sqrt{\dfrac{2}{3}x^3 + 9}$

DE: Separable Equations – Ex. 2

Solve the DE: $y' = x^2 y$, $y \neq 0$

Then, find the solution that satisfies
the initial condition, $y(0) = 5$

Rearrange	$\frac{dy}{dx} = x^2 y$ $\left(\frac{1}{y}\right) \cdot dy = x^2 \cdot dx$
Integrate	$\int \left(\frac{1}{y}\right) dy = \int x^2 \, dx$ $\ln\|y\| = \frac{1}{3}x^3 + C_1$
Solve for y Use: $e^{\ln a} = a$	$e^{\ln\|y\|} = e^{\left(\frac{1}{3}x^3 + C_1\right)}$ $\|y\| = e^{\left(\frac{1}{3}x^3\right)} \cdot e^{(C_1)}$ $\|y\| = C \cdot e^{\left(\frac{1}{3}x^3\right)}$ $y = \pm C \cdot e^{\left(\frac{1}{3}x^3\right)}$
Solve for C Use: $y(0) = 5$	$y = \pm C \cdot e^{\left(\frac{1}{3}x^3\right)}$ $5 = \pm C$
Particular Solution	$y = \pm 5\, e^{\left(\frac{1}{3}x^3\right)}$

138

DE: Separable Equations – Ex. 3	
Solve the DE: $y' = \dfrac{6x^2}{2y + \sin y}$, $y \neq 0$ To find the general solution.	
Rearrange	$\dfrac{dy}{dx} = \dfrac{6x^2}{2y + \sin y}$ $(2y + \sin y) \cdot dy = 6x^2 \cdot dx$
Integrate	$\int (2y + \sin y)\, dy = \int (6x^2)\, dx$ $y^2 - \cos y = \dfrac{6}{3} x^3 + C$ $y^2 = 2x^3 + \cos y + C$ $y = \pm\sqrt{2x^3 + \cos y + C}$
General Solution	$y = \pm\sqrt{2x^3 + \cos y + C}$
Note	$y = f(x, y)$ Because we can't get: $y = f(x)$

DE: Population Growth Model

Differential Equations (DE) Population Growth Model (Logistic Differential Equation)		
$$\frac{dP}{dt} = kP\left(1 - \frac{P}{M}\right)$$		
P = Population	$$P(t) = \frac{M}{1 + Ae^{-kt}}$$	
	$$A = \frac{M - P_0}{P_0}$$	
M = Carrying Capacity	$0 < P(0) < M$	$P' > 0$
	$M < P(0)$	$P' < 0$

DE: Population Growth Model – Ex. 1

A population is modeled by the differential

equation: $\dfrac{dP}{dt} = 1.4P\left(1 - \dfrac{P}{6200}\right)$

- For what values of P is the population increasing and decreasing?
- What are the equilibrium solutions?

Compare given equation to the general model.	$\dfrac{dP}{dt} = kP\left(1 - \dfrac{P}{M}\right)$
Identify the parts.	$M = 6200$ $k = 1.4$
Population Increases when	$0 < P < 6200$
Population Decreases when	$P > 6200$
Equilibrium When $P' = 0$	$\dfrac{dP}{dt} = 1.4P\left(1 - \dfrac{P}{6200}\right) = 0$ $P = 0$ or $P = 6200$

DE: Population Growth Model – Ex. 2

A population is modeled by the DE:

$$\frac{dP}{dt} = 0.08P\left(1 - \frac{P}{1000}\right) \quad , \quad P(0) = 300$$

Find the equation for the population, P

Identify the Parts. Compare given equation to the model.	$\frac{dP}{dt} = kP\left(1 - \frac{P}{M}\right)$ $k = 0.08$ $M = 1000$ $P_0 = 300 \quad$ Given
Find A	$A = \frac{M - P_0}{P_0}$ $A = \frac{1000 - 300}{100} = 7$
Write Eqn.	$P(t) = \frac{M}{1 + Ae^{-kt}}$ $P(t) = \frac{1000}{1 + 7e^{-.08t}}$

Parametric Equations

Curves Defined by Parametric Equations

Parametric Equations Curves Defined by Parametric Eqns.	
$x = f(t) \quad y = g(t) \qquad a \le t \le b$	
Initial Point	$\left(f(a), g(a)\right)$
Terminal Point	$\left(f(b), g(b)\right)$

Parametric Curves – Ex. 1

Sketch the curve defined by
the parametric equations:

$$x = t^2 - t \ , \quad y = t + 1, \quad 0 \le t \le 3$$

Create a T table			
	t	x	y
	0	0	1
	1	0	2
	2	2	3
	3	6	4

Plot points

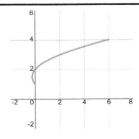

DESMOS

Code used to create graph

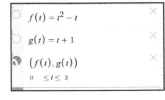

Parametric Curves – Ex. 2

Sketch the curve defined by
the parametric equations:

$$x = 3\cos t, \quad y = 3\sin t, \quad 0 \le t \le 2\pi$$

Create a T table			
	t	x	y
	0	3	0
	$\dfrac{\pi}{2}$	0	3
	π	-3	0
	$\dfrac{3\pi}{2}$	0	-3
	2π	3	0

Plot points

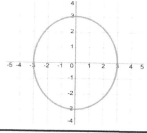

DESMOS

Code used to create graph

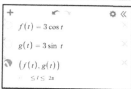

Type "pi" to get the symbol π

Parametric Curves – Ex. 3

Sketch the curve defined by
the parametric equations:

$$x = 5 + 3\cos t \ , \ y = 1 + 3\sin t \ , \ 0 \le t \le 2\pi$$

Create a T table	t	x	y
	0	8	1
	$\frac{\pi}{2}$	5	4
	π	-2	1
	$\frac{3\pi}{2}$	5	-2
	2π	8	1

Plot points	

DESMOS

Code used to create graph

$g(t) = 5 + 3\cos t$

$h(t) = 1 + 3\sin t$

$(g(t), h(t))$

$0 \le t \le 2\pi$

Type "pi" to get the symbol π

Calculus With Parametric Equations

Parametric Equations -- Calculus

$$x = f(t) \quad y = g(t) \qquad a \le t \le b$$

Slope of Tangent	$\dfrac{dy}{dx} = \dfrac{\left(\frac{dy}{dt}\right)}{\left(\frac{dx}{dt}\right)} \quad , \ \dfrac{dx}{dt} \ne 0$
Concavity	$\dfrac{d^2y}{dx^2} = \dfrac{\frac{d}{dt}\left(\frac{dy}{dx}\right)}{\left(\frac{dx}{dt}\right)}$
Area	$A = \int_a^b y \cdot dx$ $A = \int_\alpha^\beta g(t) \cdot f'(t)\, dt$
Arc Length	$L = \int_\alpha^\beta \sqrt{\left(\frac{dx}{dt}\right)^2 + \left(\frac{dy}{dt}\right)^2}\, dt$
Surface Area of Rotation	$S = \int_\alpha^\beta 2\pi r\, L\, dt$ $S = \int_\alpha^\beta 2\pi r \sqrt{\left(\frac{dx}{dt}\right)^2 + \left(\frac{dy}{dt}\right)^2}\, dt$ $r = y$ if rotated about x-axis $r = x$ if rotated about y-axis

Parametric Equations
Cycloids

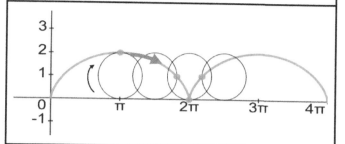

$$x = f(\theta) = r(\theta - \sin\theta)$$
$$y = g(\theta) = r(1 - \cos\theta)$$

Visualize a dot on a wheel with radius $r = 1$. As the wheel rolls along the ground, the dot marks points in a pattern of wide arches. The pattern created is a cycloid.

Parametric Curves – Calculus -- Ex. 1a
Tangents

Given the parametric equations:
$$x = t^2 \ , \quad y = t^3 - 3t$$

Find the following:

1. Sketch the curve.
2. Show the curve has two tangents at $(3,0)$.
3. Find where tangents are horiz. or vertical.
4. Find where curve is concave up or down.

Create a T table		t	x	y
		0	0	0
		1	1	-2
		2	4	2
		3	9	18
		-1	1	2

Make a sketch (1)

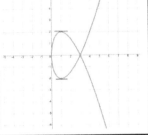

Continued ...

Parametric Curves – Calculus -- Ex. 1b
Tangents

... Continued ...

Given the parametric equations:

$$x = t^2 \, , \quad y = t^3 - 3t$$

Find the following:

2. Show the curve has two tangents at (3,0).

Find t when $(x, y) = (3, 0)$

$$x = 3 = t^2 \qquad \rightarrow \qquad t = \pm\sqrt{3}$$

$$tangent = \frac{dy}{dx} = \frac{\left(\frac{dy}{dt}\right)}{\left(\frac{dx}{dt}\right)} = \frac{3t^2 - 3}{2t}$$

$t = \sqrt{3}$ $(x, y) = (3, 0)$	$\dfrac{dy}{dx} = \dfrac{3\left(\sqrt{3}\right)^2 - 3}{2t} = \dfrac{6}{2\sqrt{3}} = \dfrac{3}{\sqrt{3}} = \sqrt{3}$ Equation: $\dfrac{\Delta y}{\Delta x} = \dfrac{(y - 0)}{(x - 3)} = \dfrac{3}{\sqrt{3}} = \sqrt{3}$ $y = \sqrt{3}\,(x - 3)$
$t = -\sqrt{3}$ $(x, y) = (3, 0)$	$\dfrac{dy}{dx} = \dfrac{3(-\sqrt{3})^2 - 3}{-2t} = -\sqrt{3}$ Equation: $\dfrac{\Delta y}{\Delta x} = \dfrac{(y - 0)}{(x - 3)} = -\sqrt{3}$ $y = -\sqrt{3}\,(x - 3)$

Continued ...

Parametric Curves – Calculus -- Ex. 1c
Tangents

... Continued ...

Given the parametric equations:
$$x = t^2 \ , \quad y = t^3 - 3t$$

Find the following:

3. Find points where tangent is horiz. or vert.

$$Slope \ = \frac{dy}{dx} = \frac{\left(\frac{dy}{dt}\right)}{\left(\frac{dx}{dt}\right)} = \frac{3t^2 - 3}{2t}$$

Horizontal Tangent when slope $= 0$	$\frac{3t^2 - 3}{2t} = 0$	
	$3t^2 - 3 = 0$	
	$3(t^2 - 1) = 0 \quad \rightarrow \quad t = \pm 1$	
	$t = 1$	$(x, y) = (1, -2)$
	$t = -1$	$(x, y) = (1, 2)$
Vertical Tangent when slope $= \infty$	$\frac{dx}{dt} = 2t = 0 \qquad \rightarrow \qquad t = 0$	
	$t = 0$	$(x, y) = (0, \ 0)$

Continued ...

Parametric Curves – Calculus -- Ex. 1d
Tangents

... Continued ...

Given the parametric equations:
$$x = t^2 \ , \quad y = t^3 - 3t$$

Find the following:

4. Find where curve is concave up or down.

1st derivative	$\dfrac{dy}{dx} = \dfrac{\left(\frac{dy}{dt}\right)}{\left(\frac{dx}{dt}\right)} = \dfrac{3t^2 - 3}{2t}$
2nd derivative	$\dfrac{d^2y}{dx^2} = \dfrac{\frac{d}{dt}\left(\frac{dy}{dx}\right)}{\left(\frac{dx}{dt}\right)} = \dfrac{\frac{d}{dt}\left(\frac{3t^2-3}{2t}\right)}{(2t)}$ $\dfrac{d^2y}{dx^2} = \dfrac{\frac{d}{dt}\left(\frac{3}{2}\right)\left(\frac{t^2-1}{t}\right)}{(2t)} = \dfrac{\frac{d}{dt}\left(\frac{3}{2}\right)\left(t - \frac{1}{t}\right)}{(2t)}$ $\dfrac{d^2y}{dx^2} = \left(\frac{3}{4}\right)\dfrac{\frac{d}{dt}(t - t^{-1})}{(t)} = \left(\frac{3}{4}\right)\dfrac{\left(1 + \frac{1}{t^2}\right)}{(t)}$ $\dfrac{d^2y}{dx^2} = \left(\frac{3}{4}\right)\left(\frac{1}{t} + \frac{1}{t^3}\right) = \left(\frac{3}{4}\right)\left(\frac{t^2 + 1}{t^3}\right)$
Conc. Up	$\dfrac{d^2y}{dx^2} = Positive \quad \rightarrow \ t > 0$
Conc. Down	$\dfrac{d^2y}{dx^2} = Negative \quad \rightarrow \ t < 0$

156

Parametric Curves – Calculus -- Ex. 2a
Tangents

Given the cycloid with $r = 1$:

$$x = r(\theta - \sin\theta)\ ,\quad y = r(1 - \cos\theta)$$
$$x = (\theta - \sin\theta)\ ,\quad y = (1 - \cos\theta)$$

Find the following:

1. Sketch the curve.

2. Find the tangent to the curve when $\theta = \dfrac{\pi}{3}$

3. Find where tangents are horiz. or vertical.

Make a sketch (1)	$f(t) = (t - \sin t)$ $g(t) = (1 - \cos t)$ $(f(t), g(t))$ $0 \le t \le 8\pi$
when $\theta = \dfrac{\pi}{3}$	$(x, y) = (\theta - \sin\theta\ , 1 - \cos\theta)$ $(x, y) = \left(\dfrac{\pi}{3} - \dfrac{\sqrt{3}}{2}, 1 - \dfrac{1}{2}\right)$ $(x, y) = \left(\dfrac{\pi}{3} - \dfrac{\sqrt{3}}{2}, \dfrac{1}{2}\right)$

Continued ...

(Stewart, Calculus Early Transcendentals, p. 650)

157

Parametric Curves – Calculus -- Ex. 2b Tangents	

Given the cycloid with $r = 1$:

$$x = (\theta - \sin\theta) \quad , \quad y = (1 - \cos\theta)$$

Find the following:

2. Find the tangent to the curve when $\theta = \frac{\pi}{3}$

Tangent when $\theta = \frac{\pi}{3}$ $(x,y) =$ $\left(\frac{\pi}{3} - \frac{\sqrt{3}}{2}, \frac{1}{2}\right)$	Slope $= \frac{dy}{dx} = \frac{\left(\frac{dy}{dt}\right)}{\left(\frac{dx}{dt}\right)} = \frac{\sin\theta}{1-\cos\theta}$ Slope $= \frac{\sin\frac{\pi}{3}}{1-\cos\frac{\pi}{3}} = \frac{\left(\frac{\sqrt{3}}{2}\right)}{\left(1-\frac{1}{2}\right)} = \sqrt{3}$
	Equation: $\frac{\Delta y}{\Delta x} = $ Slope $\frac{\left(y - \left(\frac{\pi}{3} - \frac{\sqrt{3}}{2}\right)\right)}{\left(x - \frac{1}{2}\right)} = \sqrt{3}$ $\left(y - \frac{\pi}{3} + \frac{\sqrt{3}}{2}\right) = \sqrt{3}\left(x - \frac{1}{2}\right)$ $y - \frac{\pi}{3} + \frac{\sqrt{3}}{2} = x\sqrt{3} - \frac{\sqrt{3}}{2}$ $y - \frac{\pi}{3} = x\sqrt{3} - \sqrt{3}$

| Continued ... | |

Parametric Curves – Calculus -- Ex. 2c
Tangents

Given the cycloid with $r = 1$:
$$x = (\theta - \sin\theta) \quad , \quad y = (1 - \cos\theta)$$

Find the following:

3. Find where tangents are horiz. or vertical.

Horiz. Tangent When Slope = 0	$\text{Slope} = \dfrac{dy}{dx} = \dfrac{\left(\frac{dy}{dt}\right)}{\left(\frac{dx}{dt}\right)} = \dfrac{\sin\theta}{1-\cos\theta}$	
	$\dfrac{\sin\theta}{1-\cos\theta} = 0$	Note: $\cos\theta \neq 1$ $\theta \neq 0, 2\pi, 4\pi \ldots$
	$\sin\theta = 0$	
	$\theta = \sin^{-1}(0) \quad \rightarrow \quad \theta = 0, \pi, 2\pi, \ldots$	
	$\theta = \pi, 3\pi, 5\pi \ldots$	
	$\theta = \pi$	$(x, y) = (\pi, 1)$
	$\theta = 3\pi$	$(x, y) = (3\pi, 1)$
	$\theta = 5\pi$	$(x, y) = (5\pi, 1)$

Continued ...

Parametric Curves – Calculus -- Ex. 2d
Tangents

Given the cycloid with $r = 1$:

$$x = (\theta - \sin \theta) \quad , \quad y = (1 - \cos \theta)$$

Find the following:

3. Find where tangents are horiz. or vertical.

Vertical Tangent When Slope = ∞	Slope $= \frac{dy}{dx} = \frac{\left(\frac{dy}{dt}\right)}{\left(\frac{dx}{dt}\right)} = \frac{\sin \theta}{1 - \cos \theta}$
	$\frac{\sin \theta}{1 - \cos \theta} = \infty$
	$\cos \theta = 1$
	$\theta = \cos^{-1}(1) \quad \rightarrow \quad \theta = 0, 2\pi, 4\pi, \ldots$

$\theta = 0$	$(x, y) = (0, 0)$
$\theta = 2\pi$	$(x, y) = (2\pi, 0)$
$\theta = 4\pi$	$(x, y) = (4\pi, 0)$

160

Parametric Curves – Calculus -- Ex. 3
Area

Given the cycloid:

$x = f(t) = (t - \sin t)$
$y = g(t) = (1 - \cos t)$

Find the area under
one arch of the cycloid.

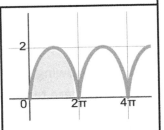

$A = \int_a^b y \cdot dx \ = \ \int_\alpha^\beta g(t) \cdot f'(t)\, dt$

$A = \int_0^{2\pi} (1 - \cos t)(1 - \cos t)\, dt$

$A = \int_0^{2\pi} (1 - 2\cos t + \cos^2 t)\, dt$

$A = \int_0^{2\pi} (1 - 2\cos t + 1 - \sin^2 t)\, dt$

$A = \int_0^{2\pi} \left(2 - 2\cos t - \frac{1}{2}(1 - \cos 2t)\right) dt$

$A = \int_0^{2\pi} \left(\frac{3}{2} - 2\cos t + \frac{1}{2}\cos 2t\right) dt$

$A = \left[\frac{3t}{2} - 2\sin t + \frac{1}{4}\sin 2t\right]_0^{2\pi}$

$A = \frac{6\pi}{2} = 3\pi$

(Stewart, Calculus Early Transcendentals, p. 651)

Parametric Curves – Calculus -- Ex. 4
Arc Length

Given the cycloid:

$x = f(t) = (t - \sin t)$

$y = g(t) = (1 - \cos t)$

Find length of one arch.

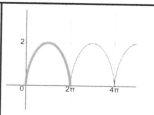

$$L = \int_\alpha^\beta \sqrt{\left(\frac{dx}{dt}\right)^2 + \left(\frac{dy}{dt}\right)^2}\ dt$$

$$L = \int_0^{2\pi} \sqrt{(1 - \cos t)^2 + (\sin t)^2}\ dt$$

$$L = \int_0^{2\pi} \sqrt{1 - 2\cos t + \cos^2 t + \sin^2 t}\ dt$$

$$L = \int_0^{2\pi} \sqrt{1 - 2\cos t + 1}\ dt$$

$$L = \int_0^{2\pi} \sqrt{2(1 - \cos t)}\ dt$$

$$\boxed{\sin^2 u = \tfrac{1}{2}(1 - \cos 2u)}$$

$$L = \int_0^{2\pi} \sqrt{4\left(\tfrac{1}{2}\right)(1 - \cos t)}\ dt$$

$$L = \int_0^{2\pi} \sqrt{4\sin^2\left(\tfrac{t}{2}\right)}\ dt = 2\int_0^{2\pi} \sin\left(\tfrac{t}{2}\right) dt$$

$$L = 2\left[-2\cos\left(\tfrac{t}{2}\right)\right]_0^{2\pi} = 2[2 + 2] = 8$$

(Stewart, Calculus Early Transcendentals, p. 653)

162

Parametric Curves – Calculus -- Ex. 4
Surface Area

Given the semi-circle
with a radius of $r = 3$

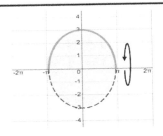

$x = f(t) = 3 \cos t$
$y = g(t) = 3 \sin t$
$0 \leq t \leq \pi$

Find the surface area of the sphere, obtained by rotating the above semi-circle about the x-axis.

$S = \int_{\alpha}^{\beta} 2\pi r \, L \, dt$

$S = \int_{\alpha}^{\beta} 2\pi y \sqrt{\left(\frac{dx}{dt}\right)^2 + \left(\frac{dy}{dt}\right)^2} \, dt$

$S = \int_{0}^{\pi} 2\pi (3 \sin t)\sqrt{(-3 \sin t)^2 + (3 \cos t)^2} \, dt$

$S = 6\pi \int_{0}^{\pi} \sin t \sqrt{9 \sin^2 t + 9 \cos^2 t} \, dt$

$S = 6\pi \sqrt{9} \int_{0}^{\pi} \sin t \sqrt{\sin^2 t + \cos^2 t} \, dt$

$S = 18\pi \int_{0}^{\pi} \sin t \, dt$

$S = 18\pi \left[- \cos t\right]_{0}^{\pi}$

$S = 18\pi[(- \cos \pi) - (- \cos 0)]$

$S = 18\pi[(1) - (-1)] = 18\pi[\, 2\,] = 36\pi$

Polar Equations

This section includes summary information about polar coordinates and curves. For more information about this topic, read my book:

Complex Numbers and Polar Curves for Pre-Calculus and Trig: With Problems and Detailed Solutions, 2023, Paulk.

Polar Coordinates and Curves

Polar Coordinates

Polar Coordinates represent the location of a point on a plane in the form (r, θ) where:

r = distance from the origin
θ = angle of direction.

A comparison with rectangular coordinates for one point is included below.

	Rectangular Example	Polar Example
Graph		
Point	$(x, y) = (4, 3)$	$(r, \theta) = (5, 37^o)$ $r = \sqrt{x^2 + y^2}$ $\theta = \tan^{-1}\left(\frac{y}{x}\right)$

Polar Coordinates – Examples

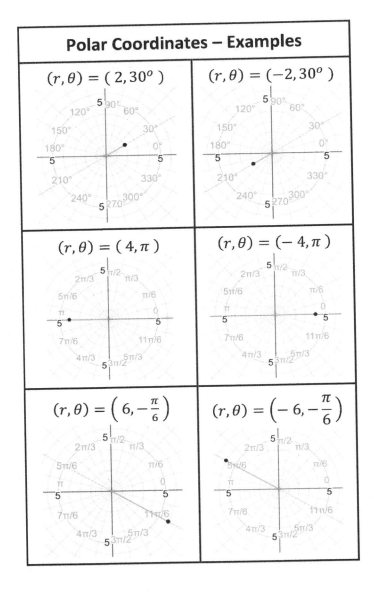

$(r, \theta) = (\,2, 30^o\,)$

$(r, \theta) = (-2, 30^o\,)$

$(r, \theta) = (\,4, \pi\,)$

$(r, \theta) = (-\,4, \pi\,)$

$(r, \theta) = \left(\,6, -\dfrac{\pi}{6}\,\right)$

$(r, \theta) = \left(-\,6, -\dfrac{\pi}{6}\,\right)$

Polar Coordinates Converting Between Rect. & Polar	
Rectangular (x, y) $\rightarrow$ Polar (r, θ)	$r = \sqrt{x^2 + y^2}$ $\theta = \tan^{-1}\left(\dfrac{y}{x}\right)$
Polar (r, θ) $\rightarrow$ Rectangular (x, y)	$x = r \cos(\theta)$ $y = r \sin(\theta)$

Note: When using a calculator to find $\theta = \tan^{-1}\left(\dfrac{y}{x}\right)$ remember that the calculator uses the default domain for tangents (Q1 & Q4). When converting points, note the quadrant in which the point is located.

	Polar Coordinates Convert Rect. To Polar Coordinates Example	
	Convert rect. coordinate $(x, y) = (-5, 3)$ to a polar coordinate.	
Original Quad.	The given point is in quadrant Q2.	
Find r	$r = \sqrt{x^2 + y^2}$ $r = \sqrt{(-5)^2 + (3)^2} = \sqrt{34} \approx 5.8$	
Find θ	$\theta = \tan^{-1}\left(\frac{y}{x}\right)$ $\theta = \tan^{-1}\left(\frac{3}{-5}\right) \approx -31^o \rightarrow Q4$ But original point is in Q2. $\theta \approx 180 - 31 = 149^o$	
Polar Coord.	$(r, \theta) = (5.8, 149^o)$	

169

Polar Coordinates **Convert Polar To Rect. Coordinates** **Example**		
Convert the polar coord. $(r,\theta) = \left(-4, \dfrac{\pi}{3}\right)$ to a rectangular coordinate.		
Original Quad.	The given point is in quadrant Q3.	
Find x	$x = r\cos(\theta)$ $x = -4\cos\left(\dfrac{\pi}{3}\right) = -4\left(\dfrac{1}{2}\right) = -2$	
Find y	$y = r\sin(\theta)$ $y = -4\sin\left(\dfrac{\pi}{3}\right) = -4\left(\dfrac{\sqrt{3}}{2}\right)$ $y = -2\sqrt{3} \approx -3.5$	
Rect. Coord	$(x,y) \approx (-2,-3.5)$ $\to Q3$	

Polar Curve

A Polar Curve is a continuous path of points on the polar coordinate system, based on a polar equation.

- In most cases, $r = f(\theta)$
  Here, r changes as the angle changes
- If $\theta = \alpha = $ constant angle
  This is a straight line, along the angle.

To graph a polar curve, for a given polar equation, create a table of values. Then graph the points on the polar coordinate system and connect the dots.

θ	$r = f(\theta)$
0	---
45°	---
90°	---
180°	---
270°	---

Polar Curve
Graphing a Polar Curve -- Example

Graph the polar curve for the given polar equation: $r = 1 + 2\cos\theta$

θ	$r = 1 + 2\cos\theta$
0	$1 + 2(1) = 3$
45°	$1 + 2\left(\frac{\sqrt{2}}{2}\right) = 1 + \sqrt{2} \approx 2.4$
90°	$1 + 2(0) = 1$
180°	$1 + 2(-1) = -1$
270°	$1 + 2(0) = 1$
360°	$1 + 2(1) = 3$

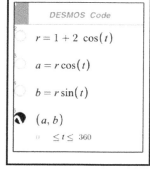

DESMOS Code

$r = 1 + 2\ \cos(t)$

$a = r\cos(t)$

$b = r\sin(t)$

(a, b)

$\ \leq t \leq\ 360$

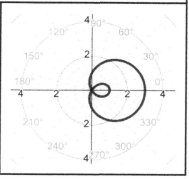

Polar Curve $r = f(\theta)$ – Examples		
Circle	$r = 2$	
Spiral	$r = \left(\dfrac{1}{4}\right)\theta$	
Rose	$r = 2\sin(5\theta)$	
Limacon	$r = 2 + \sin\theta$ Bean	
	$r = 2 + 2\sin\theta$ Cardioid	
	$r = 1 + 2\sin\theta$	
	$r = 2\sin\theta$ Circle	

Polar Curve -- Limacon Example

$$r = a + b \sin \theta \qquad a < b$$

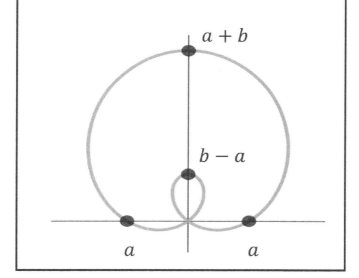

<u>Calculus With Polar Curves</u>

Polar Curves -- Calculus	
$x = r \cos \theta \qquad y = r \sin \theta \qquad r = f(\theta)$	
Slope of Tangent	$$\frac{dy}{dx} = \frac{\left(\frac{dy}{d\theta}\right)}{\left(\frac{dx}{d\theta}\right)}$$ $$\frac{dy}{dx} = \frac{\left(\frac{dr}{d\theta}\right)\sin\theta + r\cos\theta}{\left(\frac{dr}{d\theta}\right)\cos\theta - r\sin\theta}$$
Area	$$A = \int_a^b \frac{1}{2} r^2 \, d\theta$$
Arc Length	$$L = \int_a^b \sqrt{r^2 + \left(\frac{dr}{d\theta}\right)^2} \, d\theta$$

Polar Curves – Calculus -- Ex. 1
Tangents

For the Polar Curve, $r = 1 + \sin\theta$

find the slope of the tangent line when $\theta = \dfrac{\pi}{3}$

Note: The given polar curve is a cardoid.

Slope $= \dfrac{dy}{dx} = \dfrac{\left(\frac{dr}{d\theta}\right)\sin\theta + r\cos\theta}{\left(\frac{dr}{d\theta}\right)\cos\theta - r\sin\theta}$

Slope $= \dfrac{(\cos\theta)\sin\theta + (1+\sin\theta)\cos\theta}{(\cos\theta)\cos\theta - (1+\sin\theta)\sin\theta}$

Slope $= \dfrac{2\cos\theta\sin\theta + \cos\theta}{\cos^2\theta - \sin\theta - \sin^2\theta}$

Slope $= \dfrac{2\cos\left(\frac{\pi}{3}\right)\sin\left(\frac{\pi}{3}\right) + \cos\left(\frac{\pi}{3}\right)}{\cos^2\left(\frac{\pi}{3}\right) - \sin\left(\frac{\pi}{3}\right) - \sin^2\left(\frac{\pi}{3}\right)}$

Slope $= \dfrac{2\left(\frac{1}{2}\right)\left(\frac{\sqrt{3}}{2}\right) + \left(\frac{1}{2}\right)}{\left(\frac{1}{4}\right) - \left(\frac{\sqrt{3}}{2}\right) - \left(\frac{3}{4}\right)} = \dfrac{\left(\frac{2\sqrt{3}+2}{4}\right)}{\left(\frac{-2-2\sqrt{3}}{4}\right)}$

Slope $= \dfrac{2\sqrt{3}+2}{-(2\sqrt{3}+2)} = -1$

(Stewart, Calculus Early Transcendentals, p. 664)

Polar Curves – Calculus -- Ex. 2
Area

For the Polar Curve,

$r = 3\sin(2\theta)$

find the area of one leaf.

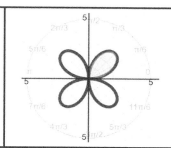

Note: First leaf occurs when $0 \le \theta \le \dfrac{\pi}{2}$

$A = \int_a^b \dfrac{1}{2} r^2 \, d\theta$

$A = \dfrac{1}{2} \int_0^{\frac{\pi}{2}} (3\sin(2\theta))^2 \, d\theta$

$A = \dfrac{9}{2} \int_0^{\frac{\pi}{2}} \sin^2(2\theta) \, d\theta$

$\qquad\qquad$ Use: $\sin^2 u = \dfrac{1}{2}(1 - \cos 2u)$

$A = \dfrac{9}{2} \int_0^{\frac{\pi}{2}} \dfrac{1}{2} (1 - \cos 4\theta) \, d\theta$

$A = \dfrac{9}{4} \left[\theta - \dfrac{1}{4}\sin 4\theta \right]_0^{\frac{\pi}{2}}$

$A = \dfrac{9}{4} \left[\left(\dfrac{\pi}{2} - 0\right) - (0) \right] = \dfrac{9\pi}{8}$

Polar Curves – Calculus -- Ex. 3a
Area

Find the area outside the cardioid $r = 1 + \cos\theta$

And inside the circle $r = 3\cos\theta$

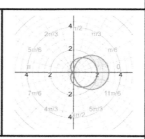

Intersections when: $1 + \cos\theta = 3\cos\theta$

$$\cos\theta = \frac{1}{2} \quad \rightarrow \quad \theta = \frac{\pi}{3}, \frac{5\pi}{3}$$

$$A = \int_a^b \left[\left(\frac{1}{2}r_2{}^2\right) - \left(\frac{1}{2}r_1{}^2\right) \right] d\theta$$

$$A = \frac{1}{2}\int_a^b \left[(3\cos\theta)^2 - (1 + \cos\theta)^2 \right] d\theta$$

$$A = \frac{1}{2}\int_a^b \left[9\cos^2\theta - 1 - 2\cos\theta - \cos^2\theta \right] d\theta$$

$$A = \frac{1}{2}\int_a^b \left[8\cos^2\theta - 1 - 2\cos\theta \right] d\theta$$

$$A = \frac{1}{2}\int_a^b \left[\frac{8}{2}(1 + \cos(2\theta)) - 1 - 2\cos\theta \right] d\theta$$

$$A = \frac{1}{2}\int_a^b \left[3 + 4\cos(2\theta) - 2\cos\theta \right] d\theta$$

$$A = \frac{1}{2} \left[3\theta + 2\sin(2\theta) - 2\sin\theta \right]_{\frac{\pi}{3}}^{\frac{5\pi}{3}}$$

Continued ...

Polar Curves – Calculus -- Ex. 3b
Area

... Continued ...

$$A = \frac{1}{2}[\, 3\theta + 2\sin(2\theta) - 2\sin\theta\,]_{\frac{\pi}{3}}^{\frac{5\pi}{3}}$$

$$A = \frac{1}{2}\left[\, 5\pi + 2\sin\frac{10\pi}{3} - 2\sin\frac{5\pi}{3}\,\right]$$

$$-\frac{1}{2}\left[\, \pi + 2\sin\frac{2\pi}{3} - 2\sin\frac{\pi}{3}\,\right]$$

$$A = \frac{1}{2}\left[\, 5\pi + 2\sin\left(\frac{6\pi}{3} + \frac{4\pi}{3}\right) - 2\sin\frac{5\pi}{3}\,\right]$$

$$-\frac{1}{2}\left[\, \pi + 2\sin\frac{2\pi}{3} - 2\sin\frac{\pi}{3}\,\right]$$

$$A = \frac{1}{2}\left[\, 5\pi + 2\sin\left(\frac{4\pi}{3}\right) - 2\sin\frac{5\pi}{3}\,\right]$$

$$-\frac{1}{2}\left[\, \pi + 2\sin\frac{2\pi}{3} - 2\sin\frac{\pi}{3}\,\right]$$

$$A = \frac{1}{2}\left[\, 5\pi + 2\left(-\frac{\sqrt{3}}{2}\right) - 2\left(-\frac{\sqrt{3}}{2}\right)\,\right]$$

$$-\frac{1}{2}\left[\, \pi + 2\left(\frac{\sqrt{3}}{2}\right) - 2\left(\frac{\sqrt{3}}{2}\right)\,\right]$$

$$A = \frac{1}{2}\,[\, 5\pi - \pi\,]$$

$$A = \frac{1}{2}[\, 4\pi\,] \;=\; 2\pi$$

Polar Curves – Calculus -- Ex. 4a
Arc Length

Find the arc length the cardioid $r = 1 + \cos\theta$	

$$L = \int_a^b \sqrt{r^2 + \left(\frac{dr}{d\theta}\right)^2}\; d\theta$$

$$L = \int_0^{2\pi} \sqrt{(1 + \cos\theta)^2 + (-\sin\theta)^2}\; d\theta$$

$$L = \int_0^{2\pi} \sqrt{(1 + 2\cos\theta + \cos^2\theta) + \sin^2\theta}\; d\theta$$

$$L = \int_0^{2\pi} \sqrt{2 + 2\cos\theta}\; d\theta$$

$$L = \int_0^{2\pi} \sqrt{2 + 2\cos\theta}\left(\frac{\sqrt{2 - 2\cos\theta}}{\sqrt{2 - 2\cos\theta}}\right) d\theta$$

$$L = \int_0^{2\pi} \left(\frac{\sqrt{(2 + 2\cos\theta)(2 - 2\cos\theta)}}{\sqrt{2 - 2\cos\theta}}\right) d\theta$$

$$L = \int_0^{2\pi} \left(\frac{\sqrt{4 - 4\cos^2\theta}}{\sqrt{2 - 2\cos\theta}}\right) d\theta$$

$$L = \int_0^{2\pi} \left(\frac{2\sqrt{1 - \cos^2\theta}}{\sqrt{2 - 2\cos\theta}}\right) d\theta$$

$$L = \int_0^{2\pi} \left(\frac{2\sqrt{\sin^2\theta}}{\sqrt{2 - 2\cos\theta}}\right) d\theta$$

Continued ...

Polar Curves – Calculus -- Ex. 4b Arc Length
... Continued ...

$$L = \int_0^{2\pi} \left(\frac{2\sqrt{\sin^2\theta}}{\sqrt{2-2\cos\theta}} \right) d\theta$$

$$L = \int_0^{2\pi} \frac{2}{\sqrt{2}} \left(\frac{|\sin\theta|}{\sqrt{1-\cos\theta}} \right) d\theta$$

$$L = \frac{4}{\sqrt{2}} \int_0^{\pi} \left(\frac{\sin\theta}{\sqrt{1-\cos\theta}} \right) d\theta \qquad \text{Use symmetry}$$

$$L = \frac{4}{\sqrt{2}} \left(\frac{\sqrt{2}}{\sqrt{2}} \right) \int_0^{\pi} \left(\frac{\sin\theta}{\sqrt{1-\cos\theta}} \right) d\theta$$

$$L = 2\sqrt{2} \int_0^{\pi} \left(\frac{\sin\theta}{\sqrt{1-\cos\theta}} \right) d\theta$$

Use u-sub $u = 1 - \cos\theta$ $du = (\sin\theta)d\theta$	$L = 2\sqrt{2} \int u^{-\frac{1}{2}} du$
	$L = 2\sqrt{2}\,(2)\,u^{\frac{1}{2}}$
	$L = 4\sqrt{2}\,\sqrt{u}$

$$L = 4\sqrt{2} \left[\sqrt{1-\cos\theta} \, \right]_0^{\pi}$$

$$L = 4\sqrt{2} \left[\sqrt{1-(-1)} - \sqrt{1-1} \, \right]$$

$$L = 4\sqrt{2} \left[\sqrt{2} - 0 \, \right]$$

$$L = 8$$

Infinite Sequences & Series

Sequences

Sequences -- Definition

A sequence is a list of numbers

in a definite order.

$$\{ a_n \} = a_1, a_2, a_3, \dots, a_n, \dots$$

A sequence $\{ a_n \}$ has the limit L where

$$\lim_{n \to \infty} a_n = L$$

If for every $\varepsilon > 0$ there is an integer N

such that If $n > N$ then $|a_n - L| < \varepsilon$

Sequences $\{a_n\} = a_1, a_2, ...$	
$\lim_{n \to \infty} a_n = L$	Convergent if the limit exists. Divergent if the limit DNE.
If $\lim_{n \to \infty} \|a_n\| = 0$ then, $\lim_{n \to \infty} a_n = 0$	
If $\lim_{n \to \infty} a_n = L$ then, $\lim_{n \to \infty} f(a_n) = f(L)$ The function must be continuous at L	
$\{r^n\}$	Convergent if: $-1 < r \leq 1$ Divergent otherwise. $\lim_{n\to\infty} r^n = \begin{cases} 0 & -1 < r < 1 \\ 1 & r = 0 \end{cases}$
$\{a_n\}$	Increasing If $a_n < a_{n+1}$ Decreasing If $a_n > a_{n+1}$ Monotonic If neither of above
$\{a_n\}$	Bounded Above If $a_n \leq M$ Bounded Below If $a_n \geq N$ Bounded If $N \leq a_n \leq M$
$\{a_n\}$	Every bonded monotonic sequence is convergent.

Sequence -- Ex. 1	
Question	Answer
Given: $a_n = \frac{n}{n+1}$ Find first four terms.	$a_n = \frac{1}{2}, \frac{2}{3}, \frac{3}{4}, \frac{4}{5}$
Given: $a_n = \cos\left(\frac{n\pi}{6}\right), n \geq 0$ Find first three terms.	$a_1 = \cos\left(\frac{\pi}{6}\right) = \frac{\sqrt{3}}{2}$ $a_2 = \cos\left(\frac{2\pi}{6}\right) = \frac{1}{2}$ $a_3 = \cos\left(\frac{3\pi}{6}\right) = 0$
Given: $a_n = 2, 4, 6, 8, \ldots$ Find the equation.	$a_n = 2n$
Given: $a_n = 5, -\frac{5}{2}, \frac{5}{3}, -\frac{5}{4}, \ldots$ Find the equation	$a_n = \left(\frac{5}{n}\right)(-1)^{n-1}$

Sequence -- Ex. 2
Find: $\lim\limits_{n\to\infty} \cos\left(\frac{\pi}{n}\right)$

Recall	If $\lim\limits_{n\to\infty} a_n = L$
	Then, $\lim\limits_{n\to\infty} f(a_n) = f(L)$
	Function must be continuous at L

$\lim\limits_{n\to\infty} \cos\left(\frac{\pi}{n}\right)$	$= \cos\left(\lim\limits_{n\to\infty}\left(\frac{\pi}{n}\right)\right)$
	$= \cos(0)$
	$= 1$

Sequence -- Ex. 3

Show that the sequence $a_n = \dfrac{n}{n^2 + 1}$

Is decreasing.

$a_{n+1} \; < \; a_n$ 　　　　　　　Must prove this.

$$\frac{n + 1}{(n + 1)^2 + 1} \; < \; \frac{n}{n^2 + 1}$$

Cross multiply

$(n + 1)(n^2 + 1) \; < \; n[(n + 1)^2 + 1\,]$

$n^3 + n + n^2 + 1 \; < \; n[\, n^2 + 2n + 1 + 1\,]$

$n^3 + n^2 + n + 1 \; < \; n[\, n^2 + 2n + 2\,]$

$n^3 + n^2 + n + 1 \; < \; n^3 + 2n^2 + 2n$

$n^2 + n + 1 \; < \; 2n^2 + 2n$

$1 \; < \; n^2 + n$

　　　　　TRUE　　　Because $n \geq 1$

Therefore: $\{\, a_n\, \}$ is decreasing

<u>Series</u>

Series	
A series is the sum of a sequence of numbers. $$s_n = \sum_{i=1}^{n} a_i = a_1 + a_2 + \cdots + a_n$$	
Convergent	$$\lim_{n \to \infty} s_n = \sum_{n=1}^{\infty} a_n = s$$ If s exists and is a real number The number s is called the sum of the series.
Divergent	If the sequence $\{s_n\}$ is divergent, then the series is also divergent.
Harmonic Series	$$\sum_{n=1}^{\infty} \left(\frac{1}{n} \right) = 1 + \frac{1}{2} + \frac{1}{3} + \cdots$$ Harmonic series is divergent.
Test for Divergence	$$\lim_{n \to \infty} a_n \neq 0$$ The series is divergent.

Series – Geometric	
Terms are multiplied by a common ratio, r $$s_n = \sum_{i=1}^{n} a \cdot r^i$$	
Infinite Geometric Series	$$s = \sum_{n=1}^{\infty} a \cdot r^n$$
Sum of Geometric Series	$$s_n = \frac{a(1-r^n)}{1-r}$$
$\lvert r \rvert \geq 1$	Geometric Series Diverges
$\lvert r \rvert < 1$	Geometric Series Converges $$s = s_\infty = \frac{a(1-r^\infty)}{1-r} = \frac{a}{1-r}$$

Series -- Ex. 1

Find the sum of the geometric series

$$5 + \frac{10}{3} + \frac{20}{9} + \frac{40}{27} + \ldots$$

Find the common ratio, r	$5r = \dfrac{10}{3}$
	$r = \dfrac{10}{3}\left(\dfrac{1}{5}\right) = \dfrac{10}{15} = \dfrac{2}{3}$
	$a_n = 5\left(\dfrac{2}{3}\right)^{n-1} = a(r)^{n-1}$
$\lvert r \rvert < 1$ $\left\lvert \dfrac{2}{3} \right\rvert < 1$	The series is convergent. $s = \dfrac{a}{1-r}$ $s = \dfrac{5}{\left(1-\dfrac{2}{3}\right)} = \dfrac{5}{\left(\dfrac{1}{3}\right)} = 15$

Series -- Ex. 2	
Is the series convergent or divergent? Given the series: $\sum_{n=1}^{\infty} 3^{2n} \cdot 7^{1-n}$	
Rewrite series in the form ar^{n-1}	$3^{2n} \cdot 7^{1-n}$ $= (3^2)^n (7)^{1-n}$ $= (9)^n (7)^{-(n-1)}$ $= \dfrac{9^n}{7^{n-1}} = \dfrac{9 \cdot 9^{n-1}}{7^{n-1}}$ $= 9 \left(\dfrac{9}{7}\right)^{(n-1)}$
$\lvert r \rvert > 1$ $\left\lvert \dfrac{9}{7} \right\rvert > 1$	The series is divergent. $r > 1$

Series -- Ex. 3	
Rewrite the number as a geometric series. $1.2\overline{34} = 1.2343434\ldots$	
Rewrite series in the form ar^{n-1}	$1.2\overline{34}$ $= 1.2 + \frac{34}{10^3} + \frac{34}{10^5} + \frac{34}{10^7} + \frac{34}{10^9}\ldots$ $= 1.2 + \sum_{n=1}^{\infty} 34 \left(\frac{1}{10}\right)^3 \left(\frac{1}{100}\right)^{n-1}$ $= 1.2 + \sum_{n=1}^{\infty} \left(\frac{34}{1000}\right)\left(\frac{1}{100}\right)^{n-1}$ Here: $a = \frac{34}{1000}$ and $r = \frac{1}{100}$
Find the sum	$S = \frac{a}{1-r} = \frac{\left(\frac{34}{1000}\right)}{1-\frac{1}{100}} = \frac{\left(\frac{34}{1000}\right)}{\left(\frac{99}{100}\right)}$ $S = \left(\frac{34}{1000}\right)\left(\frac{100}{99}\right) = \frac{34}{990}$
Write $1.2\overline{34}$ As a fraction	$1.2\overline{34} = 1.2 + s$ $= \frac{12}{10} + \frac{34}{990} = \frac{611}{495}$

194

Series -- Ex. 4

Show the series is convergent

and find the sum.   Given: $\sum_{n=1}^{\infty} \dfrac{1}{n(n+1)}$

Not a geometric series so find partial sums.

Use partial fraction decomp. to change the format.

$\dfrac{1}{n(n+1)} = \dfrac{A}{n} + \dfrac{B}{n+1}$

$1 = A(n+1) + B(n)$

$n = 0 \quad \rightarrow A = 1$

$n = -1 \quad \rightarrow B = -1$

$\sum_{n=1}^{\infty} \dfrac{1}{n(n+1)} = \sum_{n=1}^{\infty} \left(\dfrac{1}{n} - \dfrac{1}{n+1} \right)$

$s = \left(\dfrac{1}{1} - \dfrac{1}{2}\right) + \left(\dfrac{1}{2} - \dfrac{1}{3}\right) + \left(\dfrac{1}{2} - \dfrac{1}{3}\right) + \cdots \left(\dfrac{1}{n} - \dfrac{1}{n+1}\right)$

$s = 1 - \dfrac{1}{n+1}$ All terms cancel except these.

$\lim_{n \to \infty} s = 1 - \dfrac{1}{\infty} = 1 - 0 = 1$

The given series is convergent and

$$\sum_{n=1}^{\infty} \dfrac{1}{n(n+1)} = 1$$

Series -- Ex. 5
Show that the series is divergent. $$\sum_{n=1}^{\infty} \frac{n^2}{5n^2 + 6}$$
$\lim\limits_{n \to \infty} a_n = \lim\limits_{n \to \infty} \dfrac{n^2}{5n^2 + 6}$ $= \lim\limits_{n \to \infty} \dfrac{1}{5 + \frac{6}{n^2}} = \dfrac{1}{5 + 0} = \dfrac{1}{5} \ne 0$ The series diverges by the Test for Divergence

Series -- Ex. 6

Find the sum of the series.

$$\sum_{n=1}^{\infty} \frac{3}{n(n+1)} + \frac{1}{2^n}$$

$\sum_{n=1}^{\infty} \frac{3}{n(n+1)}$	$\sum_{n=1}^{\infty} \frac{1}{n(n+1)} = 1$
	From example #4.
	Therefore:
	$\sum_{n=1}^{\infty} \frac{3}{n(n+1)} = 3(1) = 3$
$\sum_{n=1}^{\infty} \frac{1}{2^n}$ Reformat to geometric form: ar^{n-1}	$\sum_{n=1}^{\infty} \frac{1}{2^n} = \sum_{n=1}^{\infty} \left(\frac{1}{2}\right)^n$ $= \sum_{n=1}^{\infty} \left(\frac{1}{2}\right)\left(\frac{1}{2}\right)^{n-1}$ Geometric series with $r = \frac{1}{2}$ $S = \frac{a}{1-r} = \frac{\left(\frac{1}{2}\right)}{\left(1-\frac{1}{2}\right)} = 1$

$$\sum_{n=1}^{\infty} \frac{3}{n(n+1)} + \frac{1}{2^n} = 3 + 1 = 4$$

Integral Test

Integral Test

If $\int_1^\infty f(x)\,dx$ is convergent,

Then $\sum_{n=1}^\infty a_n$ is convergent.

If $\int_1^\infty f(x)\,dx$ is divergent,

Then $\sum_{n=1}^\infty a_n$ is divergent.

Requirements:

- $f(x)$ is continuous, positive, and decreasing on $[1, \infty)$.
- Let $a_n = f(n)$
- Not necessary to start the series at 1.

Integral Test -- *p-series*

The p-series: $\sum_{n=1}^{\infty} \frac{1}{n^p}$ is
 	
      Convergent if $p > 1$
 	
      Divergent if $p \leq 1$

This is true, based on the Integral Test. In the section on "Improper Integrals" we learned the integral: $\int_{1}^{\infty} \frac{1}{x^p} dx$ is
 	
      Convergent if $p > 1$
 	
      Divergent if $p \leq 1$.

Integral Test -- Ex. 1
Test the series: $\sum_1^\infty \frac{2}{n^2+1}$ for conv. or diverg.

Does $f(x)$ meet the requirements? For Interval $[1, \infty)$	$f(x) = \frac{2}{n^2+1}$ • Continuous YES • Positive YES • Decreasing YES
Use the Integral Test	$\int_1^\infty f(x)\,dx = \lim_{t\to\infty} \int_1^t f(x)\,dx$

$I = \lim_{t\to\infty} \int_1^t \frac{2}{n^2+1}\,dx = \lim_{t\to\infty} 2\int_1^t \frac{1}{n^2+1}\,dx$

$I = \lim_{t\to\infty} 2\,[\tan^{-1} x]_1^t$

$I = \lim_{t\to\infty} 2\,[(\tan^{-1} t) - (\tan^{-1} 1)]$

$I = 2\,[(\tan^{-1}\infty) - (\tan^{-1} 1)]$

$I = 2\left[\frac{\pi}{2} - \frac{\pi}{4}\right] = 2\left[\frac{\pi}{4}\right] = \frac{\pi}{2}$

Conclusion: The integral is convergent so by the Integral Test, the given series is also convergent.

Integral Test -- Ex. 2
Test the series: $\sum_1^\infty \frac{1}{n^2}$ for conv. or diverg.

Note:	This is a *p-series* with $p = 2$

Conclusion: The given series is convergent, because it is a p-series, with $p > 1$.

Integral Test -- Ex. 3
Test the series: $\sum_1^\infty \frac{\ln n}{n}$ for conv. or diverg.

Does $f(x)$ meet the requirements? For Interval $[1, \infty)$	$f(x) = \frac{\ln n}{n}$ • Continuous YES • Positive YES • Decreasing YES
Use the Integral Test	$\int_1^\infty f(x)\,dx = \lim_{t\to\infty} \int_1^t f(x)\,dx$

$I = \lim_{t\to\infty} \int_1^t \frac{\ln x}{x}\,dx = \lim_{t\to\infty} \int_1^t \frac{\ln x}{x}\,dx$

Let: $u = \ln x \quad du = \left(\frac{1}{x}\right) dx$

$I = \int u\,du = \frac{u^2}{2}$

$I = \lim_{t\to\infty} \left(\frac{1}{2}\right) [\,(\ln x)^2\,]_1^t$

$I = \frac{1}{2} [\,(\infty) - (0)\,] = \infty$

Conclusion: The integral is divergent so by the Integral Test, the given series is also divergent.

Estimates of Sums

Estimating the Sum of a Series
For an infinite series: $\sum_{n=1}^{\infty} a_n$ Represented by the Integral: $\int_1^{\infty} f(x)\, dx$ $$s_n = \sum_{i=1}^{n} a_i \approx \int_1^n f(x)\, dx$$ $$s = \sum_{n=1}^{\infty} a_n \approx \int_1^{\infty} f(x)\, dx$$
If: $\quad R_n = s - s_n = $ Remainder Then: $\int_{n+1}^{\infty} f(x)\, dx \le R_n \le \int_n^{\infty} f(x)\, dx$
Add s_n to the above inequality $s_n + \int_{n+1}^{\infty} f(x)\, dx \le s \le s_n + \int_n^{\infty} f(x)\, dx$

Estimating Sums -- Ex. 1a	
Estimate the sum of the series: $\sum_1^\infty \frac{1}{n^3}$, $n = 5$	
$s_n + \int_{n+1}^\infty f(x)\,dx \leq s \leq s_n + \int_n^\infty f(x)\,dx$	
$s_5 + \int_6^\infty \left(\frac{1}{x^3}\right) dx \leq s \leq s_5 + \int_5^\infty \left(\frac{1}{x^3}\right) dx$	
$\int_5^\infty \left(\frac{1}{x^3}\right) dx$	$I = \lim\limits_{t\to\infty} \int_5^t x^{-3}\,dx$
	$I = \lim\limits_{t\to\infty} \left[\frac{x^{-2}}{-2}\right]_5^t$
	$I = \left(-\frac{1}{2}\right) \lim\limits_{t\to\infty} \left[\frac{1}{x^2}\right]_5^t$
	$I = \left(-\frac{1}{2}\right)\left[0 - \frac{1}{25}\right] = \frac{1}{50} \approx .02$
$\int_6^\infty \left(\frac{1}{x^3}\right) dx$	$I = \lim\limits_{t\to\infty} \int_6^t x^{-3}\,dx$
	$I = \left(-\frac{1}{2}\right) \lim\limits_{t\to\infty} \left[\frac{1}{x^2}\right]_6^t$
	$I = \left(-\frac{1}{2}\right)\left[0 - \frac{1}{36}\right] = \frac{1}{72} \approx .014$
Continued ...	

Estimating Sums -- Ex. 1b

Estimate the sum of the series: $\sum_1^\infty \frac{1}{n^3}$, $n = 5$

Previously found	$\int_5^\infty \left(\frac{1}{x^3}\right) dx = \frac{1}{50} \approx .02$
	$\int_6^\infty \left(\frac{1}{x^3}\right) dx = \frac{1}{72} \approx .014$

s_5	$s_5 = \sum_1^5 \frac{1}{n^3}$
	$s_5 = \frac{1}{1^3} + \frac{1}{2^3} + \frac{1}{3^3} + \frac{1}{4^3} + \frac{1}{5^3}$
	$s_5 = \frac{256103}{216000} \approx 1.186$

$$s_n + \int_{n+1}^\infty f(x)\, dx \le s \le s_n + \int_n^\infty f(x)\, dx$$

$$s_5 + \int_6^\infty \left(\frac{1}{x^3}\right) dx \le s \le s_5 + \int_5^\infty \left(\frac{1}{x^3}\right) dx$$

$$1.186 + .014 \le s \le 1.186 + .02$$
$$1.200 \le s \le 1.206$$

Approximate s to be midpoint of the interval. The error is, at most, half the interval.

$$s \approx \frac{1.200 + 1.206}{2} = 1.203 \qquad Error \le .003$$

Comparison Tests

Comparison Tests
Given: Two series with positive terms, $$\sum a_n \quad \text{and} \quad \sum b_n$$

If $\sum b_n$ is convergent and $a_n \leq b_n$ for all n	Then $\sum a_n$ Is also convergent.
If $\sum b_n$ is divergent and $a_n \geq b_n$ for all n	Then $\sum a_n$ Is also divergent.

The Limit Comparison Test	
If $\lim\limits_{n\to\infty} \dfrac{a_n}{b_n} = c$ With $c > 0$	Then both series converge or both diverge.

Comparison Tests -- Ex. 1

Does the given series converge or diverge?

$$\sum_{n=1}^{\infty} \frac{7}{2n^2 + 4n + 3}$$

Setup an inequality with a similar function.	$\dfrac{7}{2n^2 + 4n + 3} < \dfrac{7}{2n^2}$ Larger denominator on left makes the fraction smaller.

$$\frac{7}{2n^2 + 4n + 3} < \frac{7}{2n^2}$$

$$\sum_{n=1}^{\infty} \frac{7}{2n^2 + 4n + 3} < \sum_{n=1}^{\infty} \frac{7}{2n^2}$$

$$\sum_{n=1}^{\infty} \frac{7}{2n^2 + 4n + 3} < \left(\frac{7}{2}\right) \sum_{n=1}^{\infty} \frac{1}{n^2}$$

Conclusion	Right side converges by p-series with $p = 2 > 1$ So, the given series also converges by Comparison Test.

Comparison Tests -- Ex. 2

Does the given series converge or diverge?

$$\sum_{n=1}^{\infty} \frac{\ln(n)}{5n}$$

Setup an inequality with a similar function.	$\frac{\ln n}{5n} > \frac{1}{5n}$ Larger numerator on left makes the fraction larger.

$$\frac{\ln n}{5n} > \frac{1}{5n}$$

$$\sum_{n=1}^{\infty} \frac{\ln n}{5n} > \left(\frac{1}{5}\right) \sum_{n=1}^{\infty} \frac{1}{n}$$

Conclusion	Right side diverges by p-series with $p = 1 \geq 1$ So, the given series also diverges by Comparison Test.

Comparison Tests -- Ex. 3

Does the given series converge or diverge? $\sum_{n=1}^{\infty} \frac{1}{3^n - 1}$	
Use the Limit Comparison Test	$a_n = \frac{1}{3^n - 1}$ $b_n = \frac{1}{3^n} = \left(\frac{1}{3}\right)^n$ Note: b_n is a geometric series with $r = \frac{1}{3} < 1$ so it converges.

$$\lim_{n \to \infty}\left(\frac{a_n}{b_n}\right) = \lim_{n \to \infty}\left(\frac{\frac{1}{3^n - 1}}{\frac{1}{3^n}}\right) = \lim_{n \to \infty}\left(\frac{3^n}{3^n - 1}\right)$$

$$= \lim_{n \to \infty}\left(\frac{1}{1 - \frac{1}{3^n}}\right) = \frac{1}{1 - 0} = 1 > 0$$

Conclusion	b_n converges as a geometric series with $r < 1$. The given series also converges by Limit Comparison Test.

Comparison Tests -- Ex. 4

Does the given series converge or diverge?

$$\sum_{n=1}^{\infty} \frac{3n^2 + n}{\sqrt{5 + n^5}}$$

Use the Limit Comparison Test	$a_n = \dfrac{3n^2 + n}{\sqrt{5 + n^5}}$
	$b_n = \dfrac{3n^2}{n^{\frac{5}{2}}} = \dfrac{3n^{\frac{4}{2}}}{n^{\frac{5}{2}}} = \dfrac{3}{n^{\frac{1}{2}}}$
	Note: b_n is a p-series with $p = \dfrac{1}{2} < 1$ so it diverges.

$$\lim_{n\to\infty} \left(\frac{a_n}{b_n} \right) = \lim_{n\to\infty} \left(\frac{\frac{3n^2 + n}{\sqrt{5 + n^5}}}{\frac{3}{n^{\frac{1}{2}}}} \right)$$

$$= \lim_{n\to\infty} \left(\frac{n^{\frac{1}{2}} \left(3n^2 + n \right)}{3\sqrt{5 + n^5}} \right) = \lim_{n\to\infty} \left(\frac{3n^{\frac{5}{2}} + n^{\frac{3}{2}}}{3\sqrt{5 + n^5}} \right)$$

$$= \lim_{n\to\infty} \left(\frac{3 + \frac{1}{n}}{3\sqrt{\frac{5}{n^5} + 1}} \right) = \frac{3}{3\sqrt{0 + 1}} = 1 > 0$$

Conclusion	The given series also diverges by Limit Comparison Test.

Alternating Series

Alternating Series Test

An alternating series is convergent.

$$\sum_{n=1}^{\infty}(-1)^{n-1}\cdot b_n = b_1 - b_2 + b_3 - b_4 + \cdots$$

$$\text{Where } b_n > 0$$

If it meets these two requirements:

- $b_{n+1} \leq b_n$ \qquad for all n
- $\lim_{n \to \infty} b_n = 0$

Alternating Series Estimation Theorem

If an alternating series is convergent (see above)

Then

$$|R_n| = |s - s_n| \leq b_{n+1}$$

And

$$|s - s_n| \leq |s_{n+1} - s_n| = b_{n+1}$$

Alternating Series -- Ex. 1	
Does the alternating harmonic series converge? $$\sum_{n=1}^{\infty} \left(\frac{1}{n}\right) \cdot (-1)^{n-1}$$	
Check: $b_{n+1} \leq b_n$	Yes. The denominators are increasing so the terms are decreasing.
Check: $\lim_{n \to \infty} b_n = 0$	Yes. $\lim_{n \to \infty} b_n = \lim_{n \to \infty} \left(\frac{1}{n}\right) = 0$
Conclusion	The given series converges by the Alternating Series Test (AST)

Alternating Series -- Ex. 2

Does the alternating series converge or diverge?

$$\sum_{n=1}^{\infty} \left(\frac{5n}{7n-1} \right) \cdot (-1)^n$$

Check: $b_{n+1} \leq b_n$	Yes. The denominators are increasing faster so the terms are decreasing.
Check: $\lim_{n \to \infty} b_n = 0$	No. $\lim_{n \to \infty} b_n = \lim_{n \to \infty} \left(\frac{5n}{7n-1} \right) = \frac{\infty}{\infty}$ $\lim_{n \to \infty} b_n = \lim_{n \to \infty} \left(\frac{5}{7-\frac{1}{n}} \right) = \frac{5}{7}$ Can't use Alternating Series Test
Now, check the nth term.	$\lim_{n \to \infty} a_n$ $= \lim_{n \to \infty} \left(\frac{5n}{7n-1} \right)(-1)^n$ This limit DNE (It's alternating)
Conclusion	The given series diverges by the Test for Divergence.

Alternating Series -- Ex. 3	
Does the alternating series converge or diverge? $$\sum_{n=1}^{\infty} \left(\frac{n^3}{n^4 - 1} \right) \cdot (-1)^{n+1}$$	
Check: $b_{n+1} \leq b_n$	Yes. The denominator increases faster than the numerator so the terms are decreasing.
Check: $\lim_{n \to \infty} b_n = 0$	Yes. $$\lim_{n \to \infty} b_n = \lim_{n \to \infty} \left(\frac{n^3}{n^4 - 1} \right)$$ $$\lim_{n \to \infty} b_n = \lim_{n \to \infty} \left(\frac{\frac{1}{n}}{1 - \frac{1}{n^4}} \right) = \frac{0}{1} = 0$$
Conclusion	The given series is convergent by the Alternating Series Test (AST).

Alternating Series -- Ex. 4a

Find the sum of the given series correct to two decimal places. (Note: Starts with $n = 0$)

$$\sum_{n=0}^{\infty} \left(\frac{1}{n!}\right) \cdot (-1)^{n+1}$$

Check: $b_{n+1} \leq b_n$	Yes. The denominator increases faster than the numerator so the terms are decreasing.
Check: $\lim_{n \to \infty} b_n = 0$	Yes. $\lim_{n \to \infty} b_n = \lim_{n \to \infty} \left(\frac{1}{n!}\right) = 0$
Conclusion	Given series converges by AST.
Sum	$s = \frac{1}{0!} - \frac{1}{1!} + \frac{1}{2!} - \frac{1}{3!} + \frac{1}{4!} - \frac{1}{5!} + \cdots$
b_6	$b_6 = \frac{1}{6!} = \frac{1}{720} \approx .0014$
s_5	$s_5 = \frac{1}{0!} - \frac{1}{1!} + \frac{1}{2!} - \frac{1}{3!} + \frac{1}{4!} - \frac{1}{5!}$ $s_5 = \frac{11}{30} = .36667$

Continued ...

Alternating Series -- Ex. 4b

Find the sum of the given series correct to two decimal places. (Note: Starts with $n = 0$)

$$\sum_{n=0}^{\infty} \left(\frac{1}{n!} \right) \cdot (-1)^{n+1}$$

Previously we found:

- Given series is convergent.

- $b_6 = \frac{1}{6!} = \frac{1}{720} \approx .0014$

- $s_5 = \frac{11}{30} = .36667$

Alt. Series Estimation Theorem	$\|R_n\| = \|s - s_n\| \leq b_{n+1}$ $\|R_n\| \leq b_{n+1}$ (Remainder) $\|R_5\| \leq b_6 = .0014$
	The error is less than .0014 Does not affect 2nd decimal place So, we have $s \approx s_5 = \frac{11}{30} \approx .37$ $s \approx .37$ Correct to 2 decimal places

Absolute Convergence

Absolute Convergence

A series $\sum a_n$ is Absolutely Convergent

If the series of absolute values

$\sum |a_n|$ is convergent.

$\sum_{n=1}^{\infty} |a_n| = |a_1| + |a_2| + |a_3| + \cdots$

If $\sum a_n$ is a series with positive terms,

Then $|a_n| = a_n$

So, absolute convergence is

the same as convergence in this case.

If a series $\sum a_n$ is absolutely convergent,

Then it is convergent.

Conditionally Convergent

A series is conditionally convergent, if it is
convergent but not absolutely convergent

222

Absolute Convergence -- Ex. 1	
Is the series: $\sum_1^\infty \frac{1}{n^3}(-1)^{n-1}$ divergent, absolutely convergent, or convergent?	
Check $\sum \|a_n\|$	$\sum_1^\infty \left\| \frac{1}{n^3}(-1)^{n-1} \right\| = \sum_1^\infty \frac{1}{n^3}$ Which is a p-series with $p = 3$ Which is convergent.
Conclusion	Since $\sum \|a_n\|$ is convergent, The given series $\sum a_n$ is absolutely convergent. And it is also convergent.

Absolute Convergence -- Ex. 2

The alternating harmonic series is convergent.

$$\sum_1^\infty \frac{1}{n} (-1)^{n-1}$$

Is it also absolutely convergent?

Check $\sum |a_n|$	$\sum_1^\infty \left| \frac{1}{n} (-1)^{n-1} \right| = \sum_1^\infty \frac{1}{n}$ Which is a p-series with $p = 1$ Which is divergent.
Conclusion	Since $\sum |a_n|$ is not convergent, The given series $\sum a_n$ is not absolutely convergent. The given series is convergent but not absolutely convergent so it is conditionally convergent.

Absolute Convergence -- Ex. 3

Is the given series convergent,
absolutely convergent,
conditionally convergent, or divergent?

$$\sum_1^\infty \frac{\sin n}{n^2}$$

Check $\sum |a_n|$	$\sum_1^\infty \left| \frac{\sin n}{n^2} \right| = \sum_1^\infty \frac{|\sin n|}{n^2} \le \sum_1^\infty \frac{1}{n^2}$ $\sum |a_n|$ converges by comparison test with a p-series that converges with $p = 2$.
Conclusion	Since $\sum |a_n|$ converges, The given series $\sum a_n$ is absolutely convergent and also convergent.

Ratio and Root Tests

Ratio Test for $\sum a_n$	
Absolutely Convergent	$\lim\limits_{n\to\infty} \left\lvert \dfrac{a_{n+1}}{a_n} \right\rvert = L < 1$
Divergent	$\lim\limits_{n\to\infty} \left\lvert \dfrac{a_{n+1}}{a_n} \right\rvert = L > 1$ Or $\; L = \infty$
No Conclusion	$\lim\limits_{n\to\infty} \left\lvert \dfrac{a_{n+1}}{a_n} \right\rvert = L = 1$

Root Test for $\sum a_n$	
Absolutely Convergent	$\lim\limits_{n\to\infty} \sqrt[n]{\lvert a_n \rvert} = L < 1$
Divergent	$\lim\limits_{n\to\infty} \sqrt[n]{\lvert a_n \rvert} = L > 1$ Or $\; L = \infty$
No Conclusion	$\lim\limits_{n\to\infty} \sqrt[n]{\lvert a_n \rvert} = L = 1$

227

Ratio Test -- Ex. 1	
Test the series: $\sum_{1}^{\infty} \frac{n^3}{2^n}(-1)^n$ For absolute convergence.	
Setup Ratio For Ratio Test $\left\| \frac{a_{n+1}}{a_n} \right\|$	$\left\| \frac{a_{n+1}}{a_n} \right\| = \left\| \frac{\frac{(n+1)^3}{2^{n+1}}(-1)^{n+1}}{\frac{n^3}{2^n}(-1)^n} \right\|$ $= \left\| \frac{2^n}{n^3} \cdot \frac{(n+1)^3}{2^{n+1}} \cdot \frac{(-1)^{n+1}}{(-1)^n} \right\|$ $= \left\| \frac{1}{2} \cdot \frac{(n+1)^3}{n^3} \cdot (-1) \right\|$ $= \frac{1}{2} \cdot \left(\frac{n+1}{n} \right)^3$
Check $\lim_{n \to \infty} \left\| \frac{a_{n+1}}{a_n} \right\|$	$= \lim_{n \to \infty} \frac{1}{2} \left(\frac{n+1}{n} \right)^3 = \lim_{n \to \infty} \frac{1}{2} \left(\frac{1+\frac{1}{n}}{1} \right)^3$ $= \lim_{n \to \infty} \frac{1}{2} \left(\frac{1+\frac{1}{n}}{1} \right)^3 = \frac{1}{2} \left(\frac{1+0}{1} \right)^3 = \frac{1}{2}$
Conclusion	$\lim_{n \to \infty} \left\| \frac{a_{n+1}}{a_n} \right\| = \frac{1}{2} < 1$ Given series is absolutely convergent by the Ratio Test.

Ratio Test -- Ex. 2	
Test the series: $\sum_1^\infty \frac{n^n}{n!}$ for convergence.	

Setup Ratio For Ratio Test $\left\|\frac{a_{n+1}}{a_n}\right\|$	$\left\|\frac{a_{n+1}}{a_n}\right\| = \left\|\frac{(n+1)^n}{(n+1)!} \cdot \frac{n!}{n^n}\right\|$ $= \left\|\frac{(n+1)\cdot(n+1)^n}{(n+1)\cdot n!} \cdot \frac{n!}{n^n}\right\|$ $= \frac{(n+1)^n}{n^n} = \left(\frac{n+1}{n}\right)^n$
Check $\lim_{n\to\infty}\left\|\frac{a_{n+1}}{a_n}\right\|$	$= \lim_{n\to\infty}\left(\frac{n+1}{n}\right)^n$ $= \lim_{n\to\infty}\left(1+\frac{1}{n}\right)^n = e > 1$
Conclusion	$\lim_{n\to\infty}\left\|\frac{a_{n+1}}{a_n}\right\| = e > 1$ Given series is divergent by the Ratio Test.
Note	The Ratio Test works, but an easier method would be to use the Test for Divergence. $a_n = \frac{n^n}{n!} = \frac{n\cdot n\cdot n\cdot n \, \dots \, n}{1\cdot2\cdot3\cdot4 \, \dots \, n} \geq n \neq 0$

Root Test -- Ex. 1

Test the convergence of the given series.

$$\sum_1^\infty \left(\frac{5n+6}{7n+8} \right)^n$$

Take the nth root of the given series. $\sqrt[n]{\lvert a_n \rvert}$	$\sqrt[n]{\lvert a_n \rvert} = \left\lvert \frac{5n+6}{7n+8} \right\rvert = \frac{5n+6}{7n+8}$ Absolute value sign is not necessary because $a_n \geq 0$ $n \geq 1$
Check $\lim_{n\to\infty} \sqrt[n]{\lvert a_n \rvert}$	$\lim_{n\to\infty} \sqrt[n]{\lvert a_n \rvert} = \lim_{n\to\infty} \frac{5n+6}{7n+8}$ $= \lim_{n\to\infty} \frac{5+\frac{6}{n}}{7+\frac{8}{n}} = \frac{5}{7} < 1$
Conclusion	$\lim_{n\to\infty} \sqrt[n]{\lvert a_n \rvert} = \frac{5}{7} < 1$ Given series is absolutely convergent by the Root Test.

<u>Guidance for Testing Series</u>

231

Guidance for Testing Series: $\sum_{n=1}^{\infty} a_n$

a_n Example	a_n Notes	Test
$\dfrac{n-1}{2n+1}$	$a_n \to \dfrac{1}{2} \neq 0$ As $n \to \infty$	Test for Divergence
$\dfrac{\sqrt{n^3+1}}{5n^3+n^2+7}$	a_n is an algebra function of n. Compare to $b_n = \dfrac{\sqrt{n^3}}{5n^3} = \dfrac{1}{5n^{\frac{3}{2}}}$	Comparison Test b_n is a p-series
ne^{-n^2}	$\int a_n$ is easy to evaluate.	Integral Test Or Ratio Test
$\dfrac{n^3}{n^4+2}(-1)^n$	a_n is an alternating series.	Alternating Series Test
$\dfrac{3^n}{n!}$	a_n involves a factorial!	Ratio Test
$\dfrac{1}{3^n+4}$	a_n is similar to geometric series $b_n = \dfrac{1}{3^n}$	Comparison Test

Power Series

Power Series
Power Series centered at a $$\sum_{n=0}^{\infty} c_n(x-a)^n$$ $$= c_0 + c_1(x-a) + c_2(x-a)^2 + \dots$$

3 Possibilities for a Power Series	
1	Converges only when $x = a$.
2	Converges for all x.
3	Converges if $|x-a| < R$ Diverges if $|x-a| > R$

$R = $ Radius of Convergence
To find R, use the Ratio or Root Test $$\lim_{n\to\infty}\left|\frac{a_{n+1}}{a_n}\right| < 1 \quad \text{or} \quad \lim_{n\to\infty}\sqrt[n]{|a_n|} < 1$$

Radius of Convergence -- Ex. 1

For what values of x is the series convergent?
$$\sum_1^\infty n!\, x^n$$

Setup Ratio For Ratio Test $\left|\frac{a_{n+1}}{a_n}\right|$	$\left|\dfrac{a_{n+1}}{a_n}\right| = \left|\dfrac{(n+1)!\cdot x^{n+1}}{n!\cdot x^n}\right|$ $= \left|\dfrac{(n+1)\cdot n!\cdot x\cdot x^n}{n!\cdot x^n}\right|$ $= \mid (n+1)\,x\mid = (n+1)\lvert x\rvert$
Recall Ratio Test	If $\lim\limits_{n\to\infty}\left\lvert\dfrac{a_{n+1}}{a_n}\right\rvert < 1$ Then the series converges.
Find values of x that cause the series to converge.	$\lim\limits_{n\to\infty}(n+1)\lvert x\rvert\ <\ 1$ Therefore, $x=0$
Conclusion	The given series converges only when $x=0$.

Radius of Convergence -- Ex. 2

For what values of x is the series convergent?

$$\sum_1^\infty \frac{(x-5)^n}{n}$$

Setup Ratio For Ratio Test	$\left\| \frac{a_{n+1}}{a_n} \right\| = \left\| \frac{(x-5)^{n+1}}{n+1} \cdot \frac{n}{(x-5)^n} \right\|$
	$= \left\| \frac{(x-5)\,n}{n+1} \right\|$
$\left\| \frac{a_{n+1}}{a_n} \right\|$	$= \|x-5\| \left(\frac{n}{n+1} \right)$
Find values of x that cause the series to converge.	$\lim\limits_{n\to\infty} \|x-5\| \left(\frac{1}{1+\frac{1}{n}} \right) < 1$
	$\|x-5\| < 1 \qquad \rightarrow \quad R = 1$
	$-1 < x-5 < 1$
	$4 < x < 6$
Conclusion	The given series converges when: $4 < x < 6$
	The radius of convergence is: $R = 1$ Because $\|x-5\| < 1$

Radius of Convergence -- Ex. 3a

Find the radius and interval of convergence.

$$\sum_1^\infty \frac{(-4)^n \, x^n}{\sqrt{n+1}}$$

Setup Ratio For Ratio Test $\left\| \frac{a_{n+1}}{a_n} \right\|$	$\left\| \frac{a_{n+1}}{a_n} \right\| = \left\| \frac{(-4)^{n+1} x^{n+1}}{\sqrt{n+2}} \cdot \frac{\sqrt{n+1}}{(-4)^n x^n} \right\|$ $= \left\| \sqrt{\frac{n+1}{n+2}} \cdot (-4)x \right\|$ $= \sqrt{\frac{n+1}{n+2}} \, \|4x\|$
Find values of x that cause the series to converge.	$\lim_{n\to\infty} \sqrt{\frac{n+1}{n+2}} \, \|4x\| < 1$ $(1)\| 4x \| < 1$ $(4)\| x \| < 1$ $\| x \| < \frac{1}{4} \qquad \rightarrow \quad R = \frac{1}{4}$ $-\frac{1}{4} < x < \frac{1}{4}$

Continued ...

Radius of Convergence -- Ex. 3b

Find the radius and interval of convergence.

$$\sum_1^\infty \frac{(-4)^n x^n}{\sqrt{n+1}}$$

Previously we found	$\lvert x \rvert < \frac{1}{4}$ $\rightarrow$ $R = \frac{1}{4}$ $-\frac{1}{4} < x < \frac{1}{4}$

It appears the Interval of Convergence is $\left(-\frac{1}{4}, \frac{1}{4}\right)$
But, we must check the endpoints.

$x = -\frac{1}{4}$	$\sum_1^\infty \frac{(-4)^n x^n}{\sqrt{n+1}} = \sum_1^\infty \frac{(-4)^n \left(-\frac{1}{4}\right)^n}{\sqrt{n+1}}$ $= \sum_1^\infty \frac{(1)^n}{\sqrt{n+1}} = \sum_1^\infty \frac{1}{\sqrt{n+1}}$ Diverges
$x = \frac{1}{4}$	$= \sum_1^\infty \frac{(-1)^n}{\sqrt{n+1}}$ Converges
Conclusion	Interval of Converg. $\left(-\frac{1}{4}, \frac{1}{4}\right]$

Radius of Convergence -- Ex. 4a

Find the radius and interval of convergence.

$$\sum_1^\infty \frac{(-1)^n \, x^{2n}}{2^{2n} \, (n!)^2}$$

Setup Ratio For Ratio Test $\left\|\dfrac{a_{n+1}}{a_n}\right\|$	$\left\|\dfrac{a_{n+1}}{a_n}\right\|$
	$= \left\|\dfrac{(-1)^{(n+1)} \, x^{2(n+1)}}{2^{2(n+1)} \, [(n+1)!]^2} \cdot \dfrac{2^{2n} \, (n!)^2}{(-1)^n \, x^{2n}}\right\|$
	$= \left\|\dfrac{(-1) \, x^{(2n+2)}}{2^{(2n+2)} \, [(n+1)!]^2} \cdot \dfrac{2^{2n} \, (n!)^2}{x^{2n}}\right\|$
	$= \left\|\dfrac{(-1) \, x^{(2)}}{2^{(2)} \, [(n+1)!]^2} \cdot \dfrac{(n!)^2}{1}\right\|$
	$= \left\|\dfrac{(-1) \, x^2}{4 \, (n+1)! \, (n+1)!} \cdot \dfrac{n! \, n!}{1}\right\|$
	$= \left\|\dfrac{(-1) \, x^2}{4 \, (n+1) \, (n+1)} \cdot \dfrac{1}{1}\right\|$
	$= \dfrac{x^2}{4 \, (n+1)^2}$

Continued ...

Radius of Convergence -- Ex. 4b

Find the radius and interval of convergence.

$$\sum_{1}^{\infty} \frac{(-1)^n \, x^{2n}}{2^{2n} \, (n!)^2}$$

Previously we found	$\left|\frac{a_{n+1}}{a_n}\right| = \frac{x^2}{4\,(n+1)^2}$
Find values of x that cause the series to converge.	$\lim\limits_{n\to\infty} \frac{x^2}{4\,(n+1)^2} < 1$ $0 < 1 \quad \rightarrow \quad$ True for all x $\rightarrow \quad R = \infty$
Conclusion	Radius of convergence $= R = \infty$ Interval of convergence: (∞, ∞)

Representing Functions as Power Series

Functions as Power Series

$$f(x) = \frac{1}{1-x} = \sum_{n=0}^{\infty} x^n \quad , \ |x| < 1$$

$$R = 1$$

$$\frac{1}{1-x} = 1 + x + x^2 + x^3 + x^4 + \cdots$$

Differentiation and Integration , $R > 0$

$$f(x) = c_0 + c_1(x-a) + c_2(x-a)^2 + \cdots$$
$$f(x) = \sum_{n=0}^{\infty} c_n(x-a)^n$$

$\frac{d}{dx}[f(x)]$	$\frac{d}{dx}[\sum_{n=0}^{\infty} c_n(x-a)^n]$ $= \sum_{n=0}^{\infty} \frac{d}{dx}[c_n(x-a)^n]$
$\int f(x)\,dx$	$\int [\sum_{n=0}^{\infty} c_n(x-a)^n]$ $= \sum_{n=0}^{\infty} \int [c_n(x-a)^n]$

Derivative & Integral have the same R

Functions as Power Series -- Ex. 1a

Express the given function as a power series by differentiating.

$$g(x) = \frac{1}{(1-x)^2}$$

Start with something we know.	$f(x) = \frac{1}{1-x} = (1-x)^{-1}$ $f'(x) = (-1)(1-x)^{-2}(-1)$ $f'(x) = (1-x)^{-2} = \frac{1}{(1-x)^2}$
Relate the given function to the known function.	$g(x) = f'(x)$ $g(x) = \frac{1}{dx}[\, 1 + x + x^2 + x^3 \;...\,]$ $g(x) = 0 + 1 + 2x + 3x^2 + \cdots$ $g(x) = \sum_{n=1}^{\infty}(n+1)x^n$

Continued ...

Functions as Power Series -- Ex. 1b

Express the given function as a power series by differentiating.

$$g(x) = \frac{1}{(1-x)^2}$$

Previously we found	$g(x) = \sum_{n=1}^{\infty}(n+1)x^n$
EXTRA Find the radius of convergence	The derivative has the same R as the original series. $f(x) = \frac{1}{1-x} = \sum_{n=0}^{\infty}x^n$
Find R For $\sum_{n=0}^{\infty}x^n$	$\left\lvert\frac{a_{n+1}}{a_n}\right\rvert < 1$ $\left\lvert\frac{x^{n+1}}{x^n}\right\rvert < 1$ $\lvert x \rvert < 1 \qquad \rightarrow \quad R = 1$
Conclusion	The radius of convergence For the given function is $R = 1$

Functions as Power Series -- Ex. 2a	
Express the given function as a power series And find its radius of convergence. $g(x) = \ln(1+x)$	
Our known example	$f(x) = \dfrac{1}{1-x} = \sum_{n=0}^{\infty} x^n \quad, R = 1$
Try to get the given function to look like something we know.	$g'(x) = \dfrac{1}{(1+x)} = \dfrac{1}{(1-(-x))}$ $g'(x) = \sum_{n=0}^{\infty} (-x)^n$
Take integral of both sides	$\int g'(x) = \int \sum_{n=0}^{\infty} (-x)^n$ $g(x) = \int \sum_{n=0}^{\infty} (-x)^n$ $g(x) = \int \sum_{n=0}^{\infty} (x)^n (-1)^n$ $g(x) = \int [\, 1 - x + x^2 - x^3 \,]\, dx$ $g(x) = x - \dfrac{x^2}{2} + \dfrac{x^3}{3} - \cdots + C$
Continued ...	

Functions as Power Series -- Ex. 2b	
Express the given function as a power series And find its radius of convergence. $g(x) = \ln(1+x)$	
Previously we found	$g(x) = x - \dfrac{x^2}{2} + \dfrac{x^3}{3} - \cdots + C$
To find C Put $x = 0$ In original eqn. and in series.	$g(0) = \ln(1+0) = 0$ $g(0) = 0 - \dfrac{0^2}{2} + \dfrac{0^3}{3} - \cdots + C$ $0 = C$
Write $g(x)$ as a series.	$g(x) = \sum_{n=1}^{\infty} \left(\dfrac{x}{n}\right)^n (-1)^{n-1}$
R	Radius of convergence is the same as for the original series. $R = 1$

Functions as Power Series -- Ex. 3

Express the given function as a power series.

$$g(x) = \frac{x^2}{3 + x}$$

Our known example	$f(x) = \frac{1}{1 - x} = \sum_{n=0}^{\infty} x^n$
Rearrange given function to look like the known example.	$g(x) = \frac{x^2}{3 + x}$ $= \left(\frac{x^2}{3}\right)\left(\frac{1}{1 + \frac{x}{3}}\right)$ $= \left(\frac{x^2}{3}\right)\left(\frac{1}{1 - \left(-\frac{x}{3}\right)}\right)$ $= \left(\frac{x^2}{3}\right) \sum_{n=0}^{\infty} \left(-\frac{x}{3}\right)^n$ $= \left(\frac{x^2}{3}\right) \sum_{n=0}^{\infty} (-1)^n \left(\frac{x}{3}\right)^n$ $= \sum_{n=0}^{\infty} (-1)^n \cdot \frac{x^{n+2}}{3^{n+1}}$

Functions as Power Series -- Ex. 4	
Express the given function as a power series. $$g(x) = \frac{1}{1 + x^3}$$	
Our known example	$$f(x) = \frac{1}{1-x} = \sum_{n=0}^{\infty} x^n$$
Rearrange given function to look like the known example.	$$g(x) = \frac{1}{1 + x^3}$$ $$= \frac{1}{1 - (-x^3)}$$ $$= \sum_{n=0}^{\infty} (-x^3)^n$$ $$= \sum_{n=0}^{\infty} (-1)^n (x^3)^n$$ $$= \sum_{n=0}^{\infty} (-1)^n x^{3n}$$

Functions as Power Series -- Ex. 5	
Express the given function as a power series. $$g(x) = \frac{2x}{1 + x^2}$$	
Our known example	$$f(x) = \frac{1}{1-x} = \sum_{n=0}^{\infty} x^n$$
Rearrange given function to look like the known example.	$$g(x) = \frac{2x}{1 + x^2}$$ $$= (2x)\left(\frac{1}{1 + x^2}\right)$$ $$= (2x)\left(\frac{1}{1 - (-x^2)}\right)$$ $$= (2x) \sum_{n=0}^{\infty} (-x^2)^n$$ $$= (2x) \sum_{n=0}^{\infty} (-1)^n x^{2n}$$ $$= \sum_{n=0}^{\infty} 2(-1)^n x^{2n+1}$$

Taylor and Maclaurin Series

Taylor Series of $f(x)$, at a
$$f(x) \; = \; \sum_{n=0}^{\infty} \frac{f^{n}(a)}{n!} \, (x - a)^n$$
$$= f(a) + \frac{f'(a)}{1!} + \frac{f''(a)}{2!} + \frac{f'''(a)}{3!} + \; ...$$

Maclaurin Series of $f(x)$, at $a = 0$
$$f(x) \; = \; \sum_{n=0}^{\infty} \frac{f^{n}(0)}{n!} \, (x)^n$$
$$= f(0) + \frac{f'(0)}{1!} + \frac{f''(0)}{2!} + \frac{f'''(0)}{3!} + \; ...$$

Maclaurin Series – Important Examples

$f(x)$	$\sum_{n=0}^{\infty} a_n$	Expansion	R
$\dfrac{1}{1-x}$	x^n	$1 + x + x^2 + \cdots$	1
e^x	$\dfrac{x^n}{n!}$	$1 + \dfrac{x}{1!} + \dfrac{x^2}{2!} + \cdots$	∞
$\sin x$	$(-1)^n \dfrac{x^{2n+1}}{(2n+1)!}$	$x - \dfrac{x^3}{3!} + \dfrac{x^5}{5!} - \cdots$	∞
$\cos x$	$(-1)^n \dfrac{x^{2n}}{(2n)!}$	$1 - \dfrac{x^2}{2!} + \dfrac{x^4}{4!} - \cdots$	∞
$\tan^{-1} x$	$(-1)^n \dfrac{x^{2n+1}}{(2n+1)}$	$x - \dfrac{x^3}{3} + \dfrac{x^5}{5} - \cdots$	1
$\ln(1+x)$	$(-1)^n \dfrac{x^n}{n}$	$x - \dfrac{x^2}{2} + \dfrac{x^3}{3} - \cdots$	1

$$(x+1)^k = \sum_{n=0}^{\infty} \binom{k}{n} x^n$$
$$= 1 + kx + \frac{k(k-1)}{2!}x^2 + \frac{k(k-1)(k-2)}{3!}x^3 + \cdots \qquad 1$$

(Stewart, Calculus Early Transcendentals, p. 768)

Maclaurin Series -- Ex. 1a

Find the Maclaurin series for the function and its radius of convergence.

$$f(x) = \frac{1}{\sqrt{4-x}}$$

Use this example	$(1+x)^k = \sum_{n=0}^{\infty} \binom{k}{n} x^n$
Rewrite $f(x)$ In a form similar to the binomial series.	$f(x) = \frac{1}{\sqrt{4-x}} = \frac{1}{\sqrt{4\left(1-\frac{x}{4}\right)}}$ $f(x) = \frac{1}{2\sqrt{1-\frac{x}{4}}} = \frac{1}{2}\left(1-\frac{x}{4}\right)^{-\frac{1}{2}}$ $f(x) = \frac{1}{2}\left(1-\left(-\frac{x}{4}\right)\right)^{-\frac{1}{2}}$
Binomial series with $k = -\frac{1}{2}$ and Replace x with $\left(-\frac{x}{4}\right)$	$(1+x)^k = \sum_{n=0}^{\infty} \binom{k}{n} x^n$ $f(x) = \left(\frac{1}{2}\right)\sum_{n=0}^{\infty} \binom{-\frac{1}{2}}{n}\left(-\frac{x}{4}\right)^n$

Continued ...

(Stewart, Calculus Early Transcendentals, p. 767)

Maclaurin Series -- Ex. 1b

Find the Maclaurin series for the function
and its radius of convergence.

$$f(x) = \frac{1}{\sqrt{4-x}}$$

Previously we found	$f(x) = \left(\frac{1}{2}\right)\sum_{n=0}^{\infty}\binom{-\frac{1}{2}}{n}\left(-\frac{x}{4}\right)^n$

$$f(x) = \frac{1}{2}\left(1 - \frac{x}{4}\right)^{-\frac{1}{2}} = \left(\frac{1}{2}\right)\sum_{n=0}^{\infty}\binom{-\frac{1}{2}}{n}\left(-\frac{x}{4}\right)^n$$

$$= \frac{1}{2}\left[1 + \left(-\frac{1}{2}\right)\left(-\frac{x}{4}\right) + \frac{\left(-\frac{1}{2}\right)\left(-\frac{3}{2}\right)}{2!}\left(-\frac{x}{4}\right)^2\right.$$

$$+ \frac{\left(-\frac{1}{2}\right)\left(-\frac{3}{2}\right)\left(-\frac{5}{2}\right)}{3!}\left(-\frac{x}{4}\right)^3$$

$$+ \cdots + \frac{\left(-\frac{1}{2}\right)\left(-\frac{3}{2}\right)\left(-\frac{5}{2}\right)\cdots\left(-\frac{1}{2}-n+1\right)}{n!}\left(-\frac{x}{4}\right)^n + \cdots\right]$$

$$= \frac{1}{2}\left[1 + \frac{1}{8}x + \frac{1\cdot3}{2!8^3}x^2 + \cdots\right]$$

Conclusion	Series converges when $\left|-\frac{x}{4}\right| < 1$
	So, $|x| < 4$ \qquad So, $R = 4$

254

Maclaurin Series -- Ex. 2

Evaluate $\displaystyle\lim_{x\to 0}\frac{e^x-1-x}{x^2}$

Use Maclaurin series for e^x (From table)

$$e^x = \sum_{n-0}^{\infty}\frac{x^n}{n!} = 1+\frac{x}{1!}+\frac{x^2}{2!}+\cdots \quad , \ R=\infty$$

$\displaystyle\lim_{x\to 0}\frac{e^x-1-x}{x^2}$

$\displaystyle = \lim_{x\to 0}\left(\frac{1}{x^2}\right)\left[\,(\,e^x\,)-1-x\,\right]$

$\displaystyle = \lim_{x\to 0}\left(\frac{1}{x^2}\right)\left[\left(1+\frac{x}{1!}+\frac{x^2}{2!}+\cdots\right)-1-x\right]$

$\displaystyle = \lim_{x\to 0}\left(\frac{1}{x^2}\right)\left[\frac{x^2}{2!}+\frac{x^3}{3!}\cdots\right]$

$\displaystyle = \lim_{x\to 0}\left[\frac{1}{2!}+\frac{x}{3!}+\frac{x^2}{4!}+\frac{x^3}{5!}\cdots\right]$

$\displaystyle = \left[\frac{1}{2!}+\frac{0}{3!}+\frac{0^2}{4!}+\frac{0^3}{5!}\cdots\right] = \frac{1}{2}$

(Stewart, Calculus Early Transcendentals, p. 769)

255

Maclaurin Series -- Ex. 3
Find Maclaurin Series Expansion for $$f(x) = \cos(x^5)$$
Use Maclaurin series for $$\cos x = \sum_{n=0}^{\infty} \frac{(-1)^n x^{2n}}{(2n)!}$$
$$\cos(x^5) = \sum_{n=0}^{\infty} \frac{(-1)^n (x^5)^{2n}}{(2n)!}$$ $$= \sum_{n=0}^{\infty} \frac{(-1)^n x^{10n}}{(2n)!}$$

Taylor Series -- Ex. 1a

Find Taylor Series Expansion for
$$f(x) = \cos(x), \text{ near } x = 3$$

Use Taylor series expansion:

$$f(x) = \sum_{n=0}^{\infty} \frac{f^{(n)}(a)}{n!} (x-a)^n$$

Get derivatives of the function.

$f(x) = \cos x$	$\rightarrow$	$f(3) = \cos 3$
$f'(x) = -\sin x$	$\rightarrow$	$f'(3) = -\sin 3$
$f''(x) = -\cos x$	$\rightarrow$	$f''(3) = -\cos 3$
$f'''(x) = \sin x$	$\rightarrow$	$f'''(3) = \sin 3$
$f''''(x) = \cos x$	$\rightarrow$	$f''''(3) = \cos 3$

$$f(x) = \frac{\cos 3 \, (x-3)^0}{0!} + \frac{-\sin 3 \, (x-3)^1}{1!}$$
$$+ \frac{-\cos 3 \, (x-3)^2}{2!} + \frac{\sin 3 \, (x-3)^3}{3!}$$
$$+ \frac{\cos 3 \, (x-3)^4}{4!} + \frac{-\sin 3 \, (x-3)^5}{5!} \quad \cdots$$

Continued ...

Taylor Series -- Ex. 1b

Find Taylor Series Expansion for

$$f(x) = \cos(x), \text{ near } x = 3$$

Previously, we found:

$$f(x) = \frac{\cos 3 \,(x-3)^0}{0!} + \frac{-\sin 3 \,(x-3)^1}{1!}$$

$$+ \frac{-\cos 3 \,(x-3)^2}{2!} + \frac{\sin 3 \,(x-3)^3}{3!}$$

$$+ \frac{\cos 3 \,(x-3)^4}{4!} + \frac{-\sin 3 \,(x-3)^5}{5!} \quad \cdots$$

$$f(x) = \left[\frac{(-1)^n \,(\cos 3)(x-3)^{2n}}{(2n)!} \right.$$

$$\left. + \frac{(-1)^{n+1} \,(\sin 3)(x-3)^{2n+1}}{(2n+1)!} \right]$$

Tables

Integration: Logs and Exponents

Integration Table -- Exponents & Logs		
$\int du$	$=$	$u + C$
$\int u^n\, du$	$=$	$\dfrac{u^{n+1}}{n+1} + C$
$\int \dfrac{1}{u}\, du$	$=$	$\ln|u| + C$
$\int e^u\, du$	$=$	$e^u + C$
$\int a^u\, du$	$=$	$\left(\dfrac{1}{\ln a}\right) a^u + C$

Integration: Trig Functions

Integration Table -- Trig Functions	
$\int \cos u \; du$	$= \; \sin u + C$
$\int \sin u \; du$	$= \; -\cos u + C$
$\int \cot u \; du$	$= \; \ln\lvert \sin u \rvert + C$
$\int \tan u \; du$	$= \; -\ln\lvert \cos u \rvert + C$
$\int \sec u \; du \; = \; \ln\lvert \sec u + \cot u \rvert + C$	
$\int \csc u \; du \; = \; -\ln\lvert \csc u + \cot u \rvert + C$	
$\int \sec^2 u \; du$	$= \; \tan u + C$
$\int \csc^2 u \; du$	$= \; -\cot u + C$
$\int \sec u \tan u \; du \; = \; \sec u + C$	
$\int \csc u \cot u \; du \; = \; -\csc u + C$	

Integration: Inverse Trig Functions

Integration Table -- Inverse Trig Funct.
$\int \dfrac{1}{\sqrt{a^2 - u^2}}\, du \;\; = \;\; \arcsin\left(\dfrac{u}{a}\right) + C$
$\int \dfrac{1}{a^2 + u^2}\, du \;\; = \;\; \left(\dfrac{1}{a}\right) \arctan\left(\dfrac{u}{a}\right) + C$
$\int \dfrac{1}{u\sqrt{u^2 - a^2}}\, du \;\; = \;\; \arcsin\left(\dfrac{u}{a}\right) + C$

Integration: Extra Formulas

Integration Table -- Extra
$\int \ln x \ dx \ = \ x \ln\lvert x \rvert - x + C$
$\int \sec x \ dx \ = \ \ln\lvert \sec x + \cot x \rvert + C$
$\int \sec^2 x \ dx \ = \ \tan x + C$
$\int \sec^3 x \ dx$ $= \frac{1}{2} [\, \sec x \cdot \tan x$ $+ \ln\lvert \sec x + \tan x \rvert \,] \ + C$

Formula Sheets (Small Print)

It is helpful to have all the derivative, integration, and trig formulas on a few pages. Keep them handy so you can easily refer to them, during class and while doing homework.

TABLE – Trig Formulas
$sin(u \pm v) = \sin u \cos v \pm \cos u \sin v$
$cos(u \pm v) = \cos u \cos v \mp \sin u \sin v$
$tan(u \pm v) = \dfrac{\tan u \pm \tan v}{1 \mp \tan u \cdot \tan v}$

$\sin 2u = 2 \sin u \cos u$	$\tan 2u = \dfrac{2 \tan u}{1 - \tan^2 u}$

$\cos 2u = \cos^2 u - \sin^2 u = 2\cos^2 u - 1 = 1 - 2\sin^2 u$

$\sin^2 u = \dfrac{1 - \cos 2u}{2}$	$\cos^2 u = \dfrac{1 + \cos 2u}{2}$	$\tan^2 u = \dfrac{1 - \cos 2u}{1 + \cos 2u}$

$\sin u + \sin v = 2 \sin\left(\frac{u+v}{2}\right)\cos\left(\frac{u-v}{2}\right)$
$\sin u - \sin v = 2 \cos\left(\frac{u+v}{2}\right)\sin\left(\frac{u-v}{2}\right)$
$\cos u + \cos v = 2 \cos\left(\frac{u+v}{2}\right)\cos\left(\frac{u-v}{2}\right)$
$\cos u - \cos v = -2 \sin\left(\frac{u+v}{2}\right)\sin\left(\frac{u-v}{2}\right)$
$\sin u \cdot \sin v = \frac{1}{2}\left[\cos(u-v) - \cos(u+v)\right]$
$\cos u \cdot \cos v = \frac{1}{2}\left[\cos(u-v) + \cos(u+v)\right]$
$\sin u \cdot \cos v = \frac{1}{2}\left[\sin(u+v) + \sin(u-v)\right]$
$\cos u \cdot \sin v = \frac{1}{2}\left[\sin(u+v) - \sin(u-v)\right]$

TABLE - Derivatives	
$\frac{d}{dx}[cu] = cu'$	$\frac{d}{dx}[u \pm v] = u' \pm v'$
$\frac{d}{dx}[uv] = u'v + uv'$	$\frac{d}{dx}\left[\frac{u}{v}\right] = \frac{vu' - v'u}{v^2}$
$\frac{d}{dx}[c] = 0$	$\frac{d}{dx}[u^n] = n\,u^{n-1}u'$
$\frac{d}{dx}[x] = 1$	$\frac{d}{dx}[\,|u|\,] = \frac{u}{|u|}u'$
$\frac{d}{dx}[\ln u] = \frac{u'}{u}$	$\frac{d}{dx}[e^u] = e^u\,u'$
$\frac{d}{dx}[\log_a u] = \frac{u'}{(\ln a)u}$	$\frac{d}{dx}[a^u] = (\ln a)\,a^u\,u'$
$\frac{d}{dx}[\sin u] = (\cos u)\,u'$	$\frac{d}{dx}[\cos u] = -(\sin u)\,u'$
$\frac{d}{dx}[\tan u] = (\sec^2 u)\,u'$	$\frac{d}{dx}[\cot u] = -(\csc^2 u)\,u'$
$\frac{d}{dx}[\sec u] = (\sec u \cdot \tan u)\,u'$	$\frac{d}{dx}[\csc u] = -(\csc u \cot u)\,u'$
$\frac{d}{dx}[\sin^{-1} u] = \frac{u'}{\sqrt{1-u^2}}$	$\frac{d}{dx}[\cos^{-1} u] = \frac{-u'}{\sqrt{1-u^2}}$
$\frac{d}{dx}[\tan^{-1} u] = \frac{u'}{1+u^2}$	$\frac{d}{dx}[\cot^{-1} u] = \frac{-u'}{1+u^2}$
$\frac{d}{dx}[\sec^{-1} u] = \frac{u'}{|u|\sqrt{1-u^2}}$	$\frac{d}{dx}[\csc^{-1} u] = \frac{-u'}{|u|\sqrt{1-u^2}}$

268

TABLE - Integrals (Constant of Integration not included)	
$\int [\, f(u) \pm g(u)\,]\, du \;=\; \int f(u)\, du \pm \int g(u)\, du$	
$\int k \cdot f(u)\, du = k \int f(u)\, du$	$\int 1\, du = u$
$\int u^n\, du = \frac{u^{n+1}}{n+1}$	$\int \frac{1}{u}\, du = \ln\lvert u\rvert$
$\int e^u\, du = e^u$	$\int a^u\, du = \left(\frac{1}{\ln a}\right) a^u$
$\int \sin u\, du = -\cos u$	$\int \cos u\, du = \sin u$
$\int \tan u\, du = -\ln\lvert \cos u\rvert$	$\int \cot u\, du = \ln\lvert \sin u\rvert$
$\int \sec u\, du = \ln\lvert \sec u + \cot u\,\rvert$	$\int \sec^2 u\, du = \tan u$
$\int \csc u\, du = -\ln\lvert \sec u + \cot u\rvert$	$\int \csc^2 u\, du = -\cot u$
$\int \sec u \cdot \tan u\, du = \sec u$	$\int \csc u \cdot \cot u\, du = -\csc u$
$\int \frac{du}{\sqrt{a^2-u^2}} = \sin^{-1}\left(\frac{u}{a}\right)$	$\int \frac{du}{a^2+u^2} = \left(\frac{1}{a}\right)\tan^{-1}\left(\frac{u}{a}\right)$
$\int \frac{du}{u\sqrt{u^2-a^2}} = \left(\frac{1}{a}\right)\sec^{-1}\left(\frac{\lvert u\rvert}{a}\right)$	$\int \ln x\, dx = x\ln\lvert x\rvert - x$
$\int \sec u\, du \;=\; \ln\lvert \sec u + \cot u\,\rvert$	
$\int \sec^2 u\, du \;=\; \tan u$	
$\int \sec^3 u\, du \;=\; \frac{1}{2}\,[\,\sec u \cdot \tan u + \ln\lvert \sec u + \tan u\,\rvert\,]$	

<u>References</u>

- Calculus Early Transcendentals, Eighth Edition, 2015, James Stewart.

- Calculus 10e, Tenth Edition, 2014, Ron Larson, Bruce Edwards.

- Calculus Early Transcendentals Single Variable, Ninth Edition, 2009, Howard Anton, Irl Bivens, Stephen Davis.

- Differential Equations With Applications: Class Notes With Detailed Examples, 2019, Jigarkumar Patel and Kathryn Paulk.

Other Books
by Kathryn Paulk

Other Books by Kathryn Paulk

- One-Page Summaries for Algebra, Geometry, and Pre-Calculus

- Graphing Functions Using Transformations for Algebra & Pre-Calc.

- Complex Numbers and Polar Curves For Pre-Calc and Trig: With Problems and Detailed Solutions

- Calculus 1 Review in Bite-Size Pieces

- Calculus 2 Review in Bite-Size Pieces

- Calculus 3 Review in Bite-Size Pieces

- Differential Equations With Applications: Class Notes With Examples

- Discrete and Continuous Probability Distributions: A Creative Comparison (V2)

- Teach Your Child to SWIM

BIG MATH For Little Kids

Workbooks for young children
& Solution manuals for parents

- Introduction to Numbers

- Introduction to Fractions
  by Sharing Things

- Introduction to Counting and Fractions
  by Cooking Breakfast

- Learn About Fractions   *****
  by Baking Cookies

- Adding Big Numbers, Guessing Numbers
  and Secret Codes

- Learn to Graph by Riding Bikes
  on Graph Paper

Made in the USA
Middletown, DE
08 September 2024

60576574R00156